Louis Péringuey

Descriptive Catalogue of the Coleoptera of South Africa

Louis Péringuey

Descriptive Catalogue of the Coleoptera of South Africa

ISBN/EAN: 9783744756341

Printed in Europe, USA, Canada, Australia, Japan

Cover: Foto ©Thomas Meinert / pixelio.de

More available books at **www.hansebooks.com**

TRANSACTIONS

OF THE

SOUTH AFRICAN PHILOSOPHICAL SOCIETY.

VOLUME X.—PART 1.
1897.

CAPE TOWN
PUBLISHED BY THE SOCIETY

1897.

TRANSACTIONS

OF THE

SOUTH AFRICAN PHILOSOPHICAL SOCIETY.

DESCRIPTIVE CATALOGUE OF THE COLEOPTERA OF SOUTH AFRICA.—Part III.

By L. Péringuey, F.E.S., F.Z.S., &c.,

Assistant Director South African Museum.

(Read November, 1896.)

Family PAUSSIDÆ.

Buccal aperture opening downward, labrum not much developed, labial palpi three-jointed, maxillary palpi four or five-jointed, maxillæ with one inner lobe or not, short, hooked at tip, falciform or bifid, ligula horny, concave inwardly, convex outwardly, without paraglossæ, edged with bristles or setæ, both palpi and ligula hiding the mouth or not; mentum with two sharp lateral, nearly parallel lobes, median part slightly aculeate; head large and with a distinct neck (*Hylotorus* excepted); antennæ varying in number of joints from ten to two; eyes large, lateral; prothorax either entire on the upper side or nearly divided in two; elytra long, covering the abdomen; pygidium large, declivous, all the coxæ contiguous and provided with trochanters; legs short, robust, nearly always more or less compressed, sometimes broadly dilated; tarsi five-jointed; pro-, meso-, and metasternum simple; abdomen with five segments, four only of which are visible.

The habits of the singular insects included in this family are now sufficiently known.

They are myrmecophilous, and although occasionally met with in the open, the place where they should be looked for is in ants' nests in the vicinity of the heaps of larvæ brought to the surface of the formicarium for warmth. In the neighbourhood of Cape Town, where four species occur, their formicarium is generally covered with a large stone. Raffray ('Matériaux pour servir à l'étude des Paussides') says that almost all the Abyssinian species live with a very small black ant with red thorax, and that only *Paussus Curtisi* and *Hylotorus Blanchardi* are met with a small yellowish *Atta*. In South Africa I know of three kinds of ants, among which *Paussidæ* are found, and the range of two of them is very wide.

Plagiolepis fallax harbours *Pentaplatarthrus paussoides*. The formicarium of this species is occasionally very large, and the number of *P. paussoides* found in one nest has been known to exceed eighty. Three more beetles are found with this ant—*Thorictus capensis*, Pér., *Cossyphodes Bewicki*. Woll., and *Eupsalis rulsellata*.

Mr. C. N. Barker, of D'Urban, Natal, has sent me an example of *Cerapterus concolor* which he found dead, but still limp, being dragged to the nest by workers of this ant.

Herr Guienzius, who collected for a number of years round D'Urban, says, as quoted by Westwood ('Thesaurus Entomologicus Oxoniensis,' p. 73), that with few exceptions "all the specimens which he had taken were found in ants' nests, living with species which are carnivorous: *Cerapterus*, *Pleuropterus* and *Pentaplatarthrus*, with different larger species, but the true *Paussi* seem to live only with small species of ants; he had found, indeed, as many as seven distinct species of *Paussi* living with one and the same species of ants."

I am not aware that *Cerapterus* has been found, except by Guienzius in ants' nests in South Africa, and the fact of the dead specimen above mentioned being dragged to the nest does not necessarily imply that it was being brought back to its former abode, although I have related the occurrence of a somewhat similar case, but the *Paussus* (*P. Burmeisteri*) was alive. ("Notes on three Paussi," Trans. Entom. Soc. Lond., 1883, p. 138.)

Pheidole capensis, or *Pheidole punctulata*, harbours *Paussus cultratus*, *P. cucullatus*, *P. Schuckardi*, *P. Burmeisteri*, *P. Linnei*, and *P. Klugi*. I do not know of any other myrmecophilous beetle harboured by this or (?) these ants.

Acantholepis capensis harbours *Paussus lineatus*, and also two other beetles—one of the few South African *Clavigeridæ*, *Fustigerodes majusculus*, Pér.; and a *Ptinus* spec. nov.

The *Paussidæ* are occasionally found flying at the hottest time of the day, but they may be said to be crepuscular or nocturnal. The numerous examples of *Cerapterus* (two kinds) submitted to me or received at the Museum have always been caught coming to the lights in houses or at the camp fire. This is also the case with *Pleuropterus* (two kinds) and other *Paussus*; and the specimens thus caught are mostly males.

However, because these insects are found in ants' nests, it does not follow that they are of use to the ants in the way that *Aphidæ* are, and *Clavigeridæ* are said to be. I have kept in captivity for a lengthy period a good many examples of three *Paussi* (*P. Burmeisteri*, *P. Linnei*, and *P. lineatus*) as well as *Pentaplatarthrus paussoides*, and I never saw the ants attending to them at all. The *Paussi* are carnivorous and are feeding on the young larvæ, but rather than drag them away by force, the nurses prefer removing the heap of larvæ and eggs from their reach.

I have suggested (Proc. Entom. Soc., Lond., 1886, p. xxxvi) that either the crepitating power of the beetle is so well known to the ants that they make a virtue of necessity, or that they are so much accustomed to the presence of *Paussi* in their colonies through hereditary consciousness of that crepitating power that they no longer struggle against the intruders. This latter view, if correct, seems to go far in explaining why so many species of *Paussi* are found in the nests of ants belonging to the genus *Pheidole*.*

It might be objected that *Paussi* kept under unusual conditions in an artificial formicarium might adapt themselves to the conditions obtaining therein and devour the larvæ for want of the ordinary staple food, but the appetite of the examples under my observation was as keen when introduced into the formicarium as later on. I have in two instances caught *Paussus* munching away ants' larvæ in *ants' nests*.

The anatomy of the buccal organs shows, I think, unmistakably that the diet of the *Paussi* must consist of something soft; the mandibles are, it is true, long, sharp, and more or less falcate at tip, but this is not so much for the purpose of seizing the prey as of use for mating. It is by means of these organs that the male catches hold of the discoidal cavity in the prothorax of the female, and the small pads of flavescent hairs which, I believe, are *always* present in fresh examples of the genus *Paussus*, are also probably connected with copulating purposes. The jaws (*maxillæ*) are decidedly feeble, the internal lobe is seldom set with rigid spines, and they are, as a rule, deeply incised, or bifid, and eminently adapted for slow manducation accompanied by suction, such as I found to be the case with the

---

* *Paussus Favieri*, one of the two European species, and occurring in Southern France, Spain, and Algeria, is also found in the nests of a *Pheidole*, *P. megacephala*; and the other, *P. turcicus*, with *Pheidole pallidula*.

three kinds of *Paussus* which I kept in captivity. While in the act of taking food the labial palpi hang at right angles with the mouth, and no movement of the jaws is visible; in fact it was only by using a very shallow formicarium that I was able to watch the *Paussi* taking food.

It is difficult to detect the sexes of *Paussidæ* from external characters, and, with a few exceptions, dissection is the only means available. I have already stated that the males seize hold of the prothoracic cavity of the female with their jaws for mating, but this prothoracic cavity is common to both sexes in *Paussus*, and there are other genera where this prothoracic cavity is wanting, but in those all the tarsi are dilated and covered underneath with dense, short papillæ. It is well known that many male *Carabidæ* and *Hydrophilidæ* have the front, and sometimes the intermediate, tarsi provided with such cusp-like development in order to maintain the female for mating purpose, but in *Cerapterus* these papillæ occur in both sexes; of that there can be no doubt—I have proved it by dissection. It is, however, possible to recognise the female by means of the slight acumination of the median part of the apex of the pygidium. I have not yet met with the female of *Arthropterus*, but Raffray states that in this sex the tarsi are ciliate underneath, and it yet remains to be seen if in *Pleuropterus*, the other South African genus having no excavated prothorax, the tarsi of the female are papillose underneath. In *Paussus Curtisi* the antennal club is a little longer in the male than in the female, but in *P. planicollis*, an Abyssinian species, it is longer in the female than in the male.

When touched either by the hand or with a straw the *Paussidæ* crepitate, and the detonation is accompanied by the discharge of a caustic fluid which not only stains the finger as iodine or lunar caustic would, but the whole body of the insect as well as its immediate neighbourhood is covered with a bright yellow fluid, which becomes pulverulent almost immediately, and slowly disappears. Free iodine is reported to have been found in the discharge of a Javanese species, *Cerapterus quadrimaculatus*.

I am rather inclined to think that the detonation is produced by the contact of the fluid with the air, because, although expelled from behind, the anterior part of the animal is immediately covered by the yellow pulverulence, and is, therefore, in the centre of the explosion.

Raffray, in his 'Recherches anatomiques sur le *Pentaplatarthrus paussoides*,' has given a masterly account of the secreting and detonating organs of this Paussid. He finds that the organ for the secretion of the caustic fluid is really a duplicate one, one on each side of the body, independent from one another, and situated far from the rectum and anus and not connected; it opens in the upper part of the lateral angle of the pygidium, and consists of a sub-pyriform vesicle partly adhering to the tergite of the copulating clasper, and opens in a pore situated near the last stigmata, but underneath it; the sides are inflated near the opening, and have two apophyses provided with a powerful fascicle of muscles; on the top of this vesicle is a compressed, membranaceous, short tube transversely fibrous, acting as vas deferens, and ending in a large trilobate bladder of thick texture supporting a coiled vessel of moderately large diameter and consisting of two concentric tubes. This vessel, which is the secreting one, is not connected with any gland, but free and immerged in the adipose tissues.

The position of the *Paussidæ* in the systematic arrangement of the *Coleoptera* has been much discussed. Burmeister gave as his opinion that they were true *Adephaga*. It is known that among the *Carabidæ* the genus *Ozaena* has the same crepitating power as the *Paussidiæ*, and as Lacordaire says: "Not only *Ozaena* has on each elytron the peculiar lateral, posterior tubercle, but it has also another character in common, *i.e.*, the trochanters project from the internal edge of the posterior coxæ." The crepitating power of *Brachinides* is well known, and most of the *Truncatipennes* exude from the anal segment a fluid, the emission of which, however, is seldom accompanied by a detonation; *Harpalides* do the same, and under certain circumstances some of them crepitate also. Two South African species, *Stenolophus capensis* and *Acupalpus terminalis*, do occasionally detonate, and send a small column of whitish smoke when seized. I am not aware of any other Coleopterous insect included in other families that possess this detonating power, and this in itself might be an inducement to bring the *Paussidæ* in the vicinity of the *Carabidæ*, had not Raffray shown that whereas the digestive system as well as the male genital armature are like that of the Carabidous beetles, the nervous system is very dissimilar, the *Carabidæ* having twelve ganglia, of which six are abdominal, while the *Paussidæ* have seven ganglia, of which one only is abdominal.

Paussidæ can thus be considered as a very distinct family, greatly modified by the diet and habits acquired through parasitism or messmating, but having more affinities with the *Carabidæ* than with any of the other families of the order *Coleoptera*.

They occur in Europe, South America (Brazil), Asia, Ceylon, Java, Australia, Madagascar, and Africa, and now number 223 species included in 16 genera. They are represented in Africa by 7 genera, 2 of which are peculiar to this part of the world, and 92 species, while of these 6 genera and 46 species are now known to occur in South Africa. The discovery of no less than 15 new species from 1885 to date leads me to believe that this number will be much increased ultimately.

Synopsis of Genera.

Labial and maxillary palpi free, not hiding the buccal cavity.

Antennæ ten-jointed, joints not fused.

Ligula ovate, large; maxillæ with a distinct outer lobe; all tarsi with the four basal joints papillose underneath in both sexes .. .. *Cerapterus.*

Ligula very small, triangular; maxillæ without outer lobe   *Arthropterus.*

Antennæ ten-jointed, all joints but the basal one fused; maxillæ bilobate .. .. *Pleuropterus.*

Antennæ six-jointed, all joints but the basal one fused; maxillæ without outer lobe .. .. *Pentaplatarthrus.*

Labial and maxillary palpi hiding the buccal cavity.

Antennæ two-jointed.

Head with a neck, no groove in front for the reception of the antennal club .. .. *Paussus.*

Head without neck, a groove in front for the reception of the antennæ .. .. *Hylotorus.*

Gen. CERAPTERUS, Sweder.,
Vetensk. Ac. Handl., ix., 1788, p. 203.

Orthopterus, Westw.
Euthysoma, Jas. Thoms.

Head short, subelongato-quadrate, dilated behind the eyes, which are very large and prominent, narrowed into a very distinct neck; mandibles moderately long, falciform; mentum with the lateral lobes not much developed; ligula large, ovate, spatuliform, convex, finely grooved longitudinally in the centre, upper edge set with bristles; basal joint of labial palpi annuliform, second subconical, short, third much inflated, bell-shaped, truncate and hollow at tip; maxillary palpi four-jointed, apical joint as long as the two preceding ones together and subacuminate; maxillæ with an elongato-ovate membranaceous lobe, short, hooked at tip and pectinate along the inner edge; antennæ ten-jointed, basal joint subquadrangular, the others compressed, set transversely, joined in the centre only, and nearly four times as broad as long; prothorax transverse, slightly rounded laterally and nearly twice as broad as long; elytra elongato-quadrate, slightly tuberculate laterally at the apex; legs short, massive, much compressed; apical part of tibiæ hollowed so as to partly receive the tarsi; tarsi dilated, the two basal joints fused, all four thickly papillose underneath in both sexes, the fifth long, very slender, claws also long, slender and simple.

The genus is also represented in the West Coast of Africa (Sierra Leone, Guinea, Rio Grande), and in India (Pondichery), Ceylon, and Java.

Synopsis of Species.

Antennæ elongate, curving slightly outwards, transverse joints closely set together .. .. *Smithi.*
Antennæ not curved and shorter, transverse joints not closely set together .. .. *concolor.*

CERAPTERUS SMITHI, MacLeay,

Smith's Illustr. Afric. Annul., p. 74, pl. iv., fig. 1; Westw., Arcana Entom., vol. ii., p. 7, pl. xlix., fig. 4; Raffray, Matér. Etud. Pauss., pl. xvii., 7–10.

Var. *Concolor*, Schaum., Wiegm. Archiv., 1850, vol. ii., p. 169.

Chestnut brown, shining; elytra with a U-shaped flavescent band on each side of the apical part; head plane on the vertex, frontal part punctured, each puncture bearing a long seta, no impression on the vertex; basal joint of antennæ deeply punctured, densely pubescent, transverse joints pilose on both sides but more densely on the upper; prothorax twice as broad as long, a little rounded laterally, with the edges densely ciliate all round, smooth on the upper part, finely grooved longitudinally in the middle, the groove interrupted at an equal distance from base and apex, and rather plane than convex; elytra elongato-quadrate, smooth with a few short, seriated flavescent hairs on the discoidal part, denser and longer laterally; pygidium not very declivous, moderately punctured and very briefly pubescent; legs with moderately dense punctures bearing a long flavescent bristle. Length 13–15½ mm.; width 6¼–7 mm.

Hab. Transvaal (Potchefstroom, Rustenburg), Damaraland, Ovampoland, Zambezia (Buluwayo).

CERAPTERUS CONCOLOR, Westw.,

Proc. Linn. Soc., ii., 1849, p. 55.

Closely allied to the preceding species, but differentiated by the shape of the antennæ, which are not quite so long, not curved outwardly at all, and the transverse joints of which are more separated from one another than in *C. Smithi;* the head and prothorax are alike, and the elytra also, but the U-shaped apical patch is sometimes hardly distinguishable or entirely absent.

C. concolor, Sch., given in the Munich Catalogue, as well as in Raffray's list, as a variety of *C. Smithi,* might prove to be identical with the present species, but this I have not been able to ascertain. I had imagined that *C. concolor*, owing to the shorter antennæ and their straighter shape, was the female of *C. Smithi*, of which MacLeay has given an excellent figure, but on dissection I have found that these specific characters hold good in both sexes. Length 11½–14 mm.; width 5–6 mm.

Hab. Natal (D'Urban), Zambezia (Zambeze Falls, Salisbury), Zululand (Eshowe), Swazieland.

CERAPTERUS LACERATUS, Dohrn,
Stett. Ent. Zeit., 1891, p. 388.

" Not very well preserved, but yet so far recognisable that it can be plainly described, is another *Paussid*, the habitat of which, nearer than South Africa, I am not aware of. The half of the right posterior elytron and several tarsi are wanting. However, as I have waited eight years in vain for a second specimen, I describe mine as follows: *Paussus (Cerapterus) laceratus;* brownish red; elytra moderately shining and having a fulvous lunule towards the apex; eyes black; prothorax shining; elytra slightly wider than the prothorax and elongate. Length 15 mm.; width 4 mm.

"Through the shape of the well-preserved antennæ the animal is connected with *C. Smithi, Lafertei;* this confirms its African origin. It differs from these two species in having a somewhat more slender facies; the fulvous lunule at the end of the elytra, as well as the characteristic antennæ and tarsi, show the animal to be unmistakably a *Cerapterus*. The species of that genus in my collection (*C. Smithi, concolor, Lafertei*) are all dark brown, nearly black, but *C. laceratus* is light reddish brown; this and a somewhat narrower prothorax give it a more slender appearance. I have no doubt that it is specifically different."

GEN. ARTHROPTERUS, MacLeay,
Smith's Illustr. Afric. Annul., 1838, p. 75.

Head and antennæ as in *Cerapterus;* maxillæ without outer lobe, not hooked at tip but with six sharp teeth along the inner edge; ligula very small, triangular; legs compressed, anterior tarsi of male not much dilated, papillose underneath, posterior and intermediate slightly less papillose underneath and more villose laterally, those of the female ciliate underneath; prothorax somewhat cordiform; head with two round depressions on the vertex.

This genus has several representatives in Australia, and another species is recorded from Abyssinia.

ARTHROPTERUS KIRBYI, Westw.,
Proc. Ent. Soc., 1864, p. 189; Thesaur. Entom. Oxon., p. 79, pl. xvi., fig. 1.

Chestnut brown, glabrous, shining; antennæ long, the transverse joints of equal width, except the second one, closely set, and hardly pubescent; head with two deep impressions on the vertex close to the eyes; prothorax subelongato-quadrate, marginate, slightly sinuate above the basal angle and with a lateral impression on each side, smooth, not fringed with hairs, and with a narrow median longitudinal groove reaching neither base nor apex; elytra straight laterally and not much broader at the base than the prothorax, but gradually ampliated till the apex, where they are not truncate, but gradually rounded off, quite smooth and hairless, with an elongated flavescent lateral patch above the subtuberculated fold, the posterior margin is also narrowly flavescent, and the suture has a very narrow line of the same colour ascending from the apex to a little past the median part; the tibiæ are moderately broad and compressed, and the anterior ones a little arcuate inwardly.

The only specimen that I have seen (male) differs a little from the figure given by Westwood. The two impressions on the vertex are not in front of the eyes, but at about the median part and close to them; the prothorax is of the same colour as the head and elytra, and not flavescent, and has no trace of two discoidal round impressions, and the elytra are not truncate behind, but sloping and gradually rounded from the posterior subtuberculated part of the fold. There is little doubt however that the two specimens belong to one species. Length 7 mm.; width 3 mm.

Hab. Natal (D'Urban).

GEN. PLEUROPTERUS, Westw.,
Trans. Linn. Soc., xviii., 1841, p. 585.

Heteropaussus, Jas. Thoms.

Head elongato-quadrate, not dilated behind the eyes, but slightly constricted there and prolonged into a neck not much narrower than the head; maxillary palpi quadri-articulate, second joint as long as the two following, apical one subacuminate; maxillæ with a very short, subrudimentary outer lobe, inner lobe hooked at tip and pluridentate along the inner edge; ligula convex, broad, ovate, convex outwardly and carinate in the centre; labial palpi triarticulate, last joint swollen, subelongato-ovate and subacuminate; labrum sharply triangular; antennæ compressed, ten-jointed, basal joint subelongato-quadrate, the other joints transverse, partly or completely fused together; prothorax transverse with the sides recurved; elytra elongate, parallel, costate or partly costate, and projecting a little beyond the pygidium; legs slender, the tibiæ especially; intermediate coxæ irregular in shape and femora compressed; tarsi long; in the male the four basal joints of every tarsi are thickly papillose underneath, the first basal joint is small, the second is dilated, elongato-quadrate, and longer than the following two together; posterior tibiæ of male slightly sinuate. I have not seen any female example as yet.

Besides the two South African species, two more are recorded from Africa (Senegal, Congo), and one from Ceylon.

Synopsis of Species.

Elytra with four rounded costæ on each side, testaceous with two discoidal longitudinal bands on each side .. .. *alternans*.
Elytra with a short basal costa, black, with the suture, a short basal band, a post-median sinuate patch and the posterior margin light testaceous .. .. *hastatus*.

PLEUROPTERUS ALTERNANS, Westw.,

Proc. Linn. Soc., ii., 1849, p. 56; Thesaur. Entom. Oxon., p. 74, pl. xvi., fig. 2.

Head, antennæ, prothorax, and legs reddish brown; elytra testaceous and having on each side two broad, longitudinal dorsal black bands uniting above the apex; head deeply punctured and densely bristly; antennæ with short, squamiform hairs, very closely set together and united by a broad node, basal joint and also the long outer spur of the second joint slightly pilose; prothorax nearly twice as broad as long, almost straight in front, sinuate laterally with the posterior angle penicillate, depressed on the lateral part of the disk, subgibbose in the middle, finely grooved longitudinally and with the posterior median part excavate, with the sides of the excavation produced in a triangular projection rounded at tip; elytra with the shoulders rounded, elongate, parallel, and having on each side six distinct costæ, the two dorsal ones of which are more raised than the other two, the second dorsal costa is the only one that reaches the base; lateral and posterior margins with a few rigid setæ; pygidium glabrous with a fringe of very short bristles; femora nearly glabrous; tibiæ setulose. Length 9–10 mm.; width 4–4¼ mm.

Hab. Natal (D'Urban, Eshowe). Is said to have been captured in Mozambique and at Lake N'Gami.

P. HASTATUS, Westw.,
Proc. Linn. Soc., 1849, p. 57 ; Thesaur. Entom. Oxon., p. 74,
pl. xvi., fig. 3.

Palpi, antennæ, and legs dark chestnut brown ; head and prothorax varying from dark brown to piceous red ; elytra black with a basal dorsal elongated flavescent patch, and a diagonal post-median transverse one reaching from one median part of the disk to the other, and much narrowed on each side of the suture, the posterior margin has also a moderately broad band, which narrows and ascends along the suture as far as the transverse dorsal band ; head very rugose, pilose, and with two round tubercles at the apical part of the vertex ; antennæ briefly pilose, with the hairs longer on the basal joint and on the inner and outer angles of the second one, joints very closely set, and seemingly fused altogether ; prothorax twice as broad as long, slightly sinuate in front with the sides rounded, recurved, sinuate in the posterior part, with the basal angle slightly sloping and penicillate, discoidal part moderately plane, and with a narrow longitudinal groove ; sides depressed ; posterior part very broadly excavate for nearly two-thirds of the width, and with a small tubercular projection at each end ; elytra with the shoulders a little rounded, thickly but briefly pubescent, very faintly striate, but with a very well-developed basal, round costa, reaching only one-sixth of the length ; pygidium punctulate, briefly pubescent, and with a thick fringe of short hairs on the margins. Length $10\frac{1}{2}$–$10\frac{3}{4}$ mm. ; width $4\frac{1}{2}$ mm.

Hab. Natal (D'Urban, Upper Districts). Female unknown.

GEN. PENTAPLATARTHRUS, Westw.,
Trans. Linn. Soc., xvi., 1833, p. 616 ; Raffray, Matér. Etud.
Pauss., pl. xvii., figs. 17, 18, 21.

Head elongato-quadrate, surrounding the eyes behind ; neck very short ; maxillary palpi quadri-articulate, basal one conical, shorter than the other three, second as long, thicker than the apical one, which is subacuminate ; maxillæ broad, short, not hooked at tip, but fringed with bristles in the inner and outer edges, and without any trace of outer lobe ; labial palpi triarticulate, apical joint as long as the two preceding ones, swollen truncate and hollowed at tip ; labrum in the shape of a broad triangle ; mandibles straight at the base as far as the median part, and diagonal from there to the apex, which is very sharp ; antennæ six-jointed, basal joint irregular, a little incurved, the other five joints flat, transverse, fused together, second joint one-fourth the length of the third ; prothorax spinose laterally, tuberculate in the anterior part, excavate behind; elytra elongato-quadrate, with the shoulders not rounded, truncate behind; pygidium perpendicular, triangular, deeply excavate on the dorsal part, and with the declivity convex, the upper margin incised in the middle, with the incision filled with a yellow pubescence; femora and tibiæ compressed, dilated; tarsi neither dilated nor papillose.

Two species of this genus have been lately described, from the East Coast of Africa (Dar-es-Salaam and Somaliland), which, judging from the description, are very closely allied to the two South African species. I have also seen a very large example from the Zanzibar mainland, in the collection of Mr. R. Oberthur.

Synopsis of Species.

Elytra totally chestnut brown .. .. *paussoides*.
Elytra with a broad median transverse infuscate band, and a narrower supra-apical one .. .. *natalensis*.

PENTAPLATARTHRUS PAUSSOIDES, Westw.,

Trans. Linn. Soc., xvi., 1833, p. 619, pl. xxxiii., figs. 1–14; Arcana Entom., ii., p. 38, pl. lviii., fig. 2.

Chestnut brown, shining; antennæ squamose; head deeply punctured, with the punctures squamose, elongato-quadrate and with a broad, deep impression on the vertex, neck not narrower than the head, and more closely punctured; prothorax subcordate, with the anterior part ridged, the ridge high, broadly incised in the centre, with the lobes divaricating, and also on each side where the incised part is developed in a lateral blunt spine; the anterior part of the ridge has two deep impressions; the median part has a deep excavation edged on each side by a narrow, rounded ridge, and produced up to the base in a deep, broad groove; it is nearly glabrous; elytra elongato-quadrate, slightly rounded at the humeral angle, truncate behind, roughly and irregularly punctured, and very briefly pubescent; in the outer and posterior margins there is a series of short setæ, and two at the tip of the posterior declivity; legs briefly setulose. Length 6–7½ mm.; width 2½–3¼ mm.

Hab. Cape Colony (Cape Town, Stellenbosch, Carnarvon, Beaufort West).

PENTAPLATARTHRUS NATALENSIS, Westw.,

Proc. Linn. Soc., ii., 1849, p. 57.

In shape and sculpture *P. natalensis* can hardly be distinguished from *P. paussoides*, but it is always larger, and the elytra, instead of being of a uniform colour all over, are distinctly redder, and have a broad median transverse dark band, sometimes piceous black, as well as a narrower one edging the apex; owing to the larger size, the punctures on the elytra are also deeper and coarser, and so far as I know *P. natalensis* does not occur in the Cape Colony proper, whereas I have not recorded *P. paussoides* from anywhere else. Length $8\frac{1}{2}$–$9\frac{1}{2}$ mm.; width $3\frac{1}{2}$–4 mm.

Hab. Free State (Vaal River), Transvaal (Potchefstroom), Ovampoland (Okovango River).

Gen. PAUSSUS, Linn.,
Bigæ Insect., Upsal, 1775, p. 7.

Maxillary palpi four-jointed, the second longer and wider than the others, labial palpi three-jointed, the apical joint longer than the others and more or less acuminate at tip; maxillæ without internal lobe, short and bifid; ligula transverse, sometimes slightly sinuate, always setose at tip, concave inwardly, convex outwardly, and covering, with the palpi, the buccal cavity; head declivous in front, more or less elongato-quadrate, dilated behind the eyes, and narrowed into a distinct neck; vertex with either a conical, sometimes penicillated, spine, ridges, depressions, or prominences bearing two small fossæ; eyes reniform or oval, with the posterior part of the head edging the eye, projecting often and sometimes aculeate; antennæ two-jointed, the basal joint thick, more or less quadrate, the second one varying much in shape; prothorax either transverse or cordiform, in which case it is incised laterally and impressed transversely, or made bipartite by a deep, transverse, sinuous groove, but having always on each side a small patch of dense, short, flavescent hairs; elytra elongato-quadrate, parallel or subparallel, not much convex, covering the whole abdomen except the pygidium, and having on each side of the apical angle a small but very distinct ridged tubercle; they are more or less deeply punctulate and pubescent, the pubescence being sometimes reduced to squamiform scattered hairs, but are not striate; legs short, compressed, bristly, setulose or squamose; femora claviform, subclaviform or compressed; tibiæ either subcylindrical, moderately compressed or dilated; tarsi five-jointed, joints of anterior pair bristly underneath in both sexes.

From the diversity of characteristics given in this diagnosis, it is seen how difficult of arrangement the species of *Paussus* will prove to be. The extraordinary shape of the second joint, or antennal club, is probably the means of identifying the species. No antennal clubs are exactly alike, and they vary in shape from a more or less regularly lenticular, round, oval, oblong, laminiform to cylindrical; the outer margin is often ampliate, excavate, or grooved, in which case the edges of the hollowed margin are more or less setigerous, and they have also on either the upper declivity, or on both sides, transverse striæ, which might perhaps imply that the articulations have become fused together.

Few South African *Paussus* can be said to have a close ally, excepting *P. lineatus* and *P. Afzelii*, which are however very distinct; *P. Schaumi* and *P. Germari*, which will probably prove to be identical; and *P. cucullatus* and *P. ruber*. If the shape of the maxillary palpi were taken into consideration, the South African species could be divided in eight groups:—

1. Second joint of maxillary palpi about equal to or a little shorter than the two following, subcylindrical, curving a little outwardly, and tapering slightly from base to apex: *P. Humboldti, damarinus, mimus, spinicoxis, propinquus, rusticus, manicanus, fallax, Bohemani*.

2. Second joint longer than the other two following: *P. signatipennis*.

3. Second joint shorter than the two following; apical joint of maxillæ long, and nearly tapering from base to apex: *P. cultratus*.

4. Second joint as long as the other two, but twice as wide; last joint of maxillary palpi short, elongato-ovate: *P. lineatus, Afzelii*.

5. Second joint elongato-quadrate, or curved outwardly from base to middle, and diagonal from there to apex, truncate at tip, where it is broader than the two following: *P. cylindricornis, Schuckardi, Curtisi*.

6. Second joint nearly straight inwardly, much swollen and rounded outwardly, the two joints following small, narrow: *P. Klugi, cucullatus, Burchellianus, ruber, cochlearius, viator, Linnei Burmeisteri, Marshalli*.*

7. Second joint broadly inflated, nearly hexagonal, irregularly rounded outwardly and deeply incised at base, inwardly: *P. granulatus*.

8. Second joint broadly quadrate: *P. Schaumi, Germari*.

Where the second joint is inflated it is slightly convex outwardly and concave inwardly.

To the first group belong all species with a conical spine on the vertex of the head, whether with bipartite prothorax or not. The second contains only one species, as does the third; but these species are strikingly distinct, as is also the case in groups seven and eight;

* In *P. Marshalli* the second joint is broader and more quadrate inwardly, and the two joints following are longer.

in group five are three species with cylindrical antennal club, but there is a little difference between *Curtisi* and *cylindricornis*, but they agree in having the second joint broader at tip than the third, and truncate, while the sixth group includes all species with bipartite prothorax, the anterior part of which is ridged, lenticular, or in the shape of a broadly truncate one.

This arrangement is, however, somewhat artificial, but in order to make the identification easier I have adopted another one which is more artificial still, but which, I hope, will facilitate the identification.

Synopsis of Species.

I.

Head with a conical spine on vertex.

a.

Club of antennæ more or less ovate, but always sinuate in the posterior margin, thickened in the centre, and a little longer than broad.

b.

Prothorax impressed transversely and constricted laterally in the centre, but with the anterior part not much more raised than the posterior.

c.

Posterior tibiæ compressed but slender, femora moderately clavate.

Club of antennæ very thick, hardly marginate behind, and with four longitudinal, shallow striæ on the upper part of the outer declivity; prothorax quadrate, divided in two parts of nearly equal size by a deep, transverse impression .. .. .. *humboldti.*

Club much thickened in the posterior part, bisinuate outwardly and having four longitudinal striæ on the upper and under parts of the outer declivity and indenting the edges of the outer margin, which is slightly scooped; prothorax rounded laterally, constricted behind and deeply impressed in the middle *damarinus.*

Club subelongate, not very thick and of nearly equal thickness, anterior margin nearly straight .. .. *mimus.*

Club subelongate, with five deep striæ on the outer declivity, outer margin deeply grooved, both edges with five impressions .. *dohrni.*

Club subelongate, acutely marginate all round and having on both sides of the outer declivity four longitudinal striæ not indenting the outer margin; prothorax with the anterior angles rounded, the sides nearly straight, and with a deep median cavity .. .. *spinicoxis.*

Club subelongate, not much convex, sharply carinate all round and without any striation in the outer declivity; median part of prothorax deeply impressed .. .. *propinquus.*

Club lenticular, carinate all round, outer margin with three very small indentations in the middle; prothorax with a shallow median impression, and very deeply constricted laterally .. .. *rusticus.*

Club of antennæ globose, carinate all round, and without any striation on the outer declivity; prothorax nearly parallel and with a deep, transverse impression .. .. *arduus*.

c c.

Femora very slender at base and strongly clavate at apex; tibiæ incurved.

b b.

Prothorax bipartite, anterior part perpendicular and much higher than the posterior.

Club thick, rounded, sharply marginate inwardly, produced in a sharp, recurved spine at the tip, posterior margin nearly straight, slightly grooved, upper and under parts of the outer declivity with four shallow striæ, the intervals of which dent slightly both edges .. .. *bohemani*.

Club thick, rounded, sharply marginate inwardly, sinuate at tip, outer part sinuate at apex and base with the basal angle long and sharp, margin slightly grooved, and with two faint impressions in the groove, no striæ on the outer declivity .. .. *fallax*.

II.

Head without a vertical, conical spine; pygidium not bristly.

a.

Club of antennæ thickened, longer than broad, outer margin neither grooved nor scooped.

b.

Prothorax impressed transversely, anterior part not much more raised than the posterior.

c.

Posterior tibiæ compressed but slender.

Club oblong, outer margin not sinuate; prothorax constricted laterally past the middle, and with a deep, median, transverse impression .. .. *manicanus*.

Club elongato-ovate, outer margin bisinuate and incised at the base above the basal outer spur; prothorax constricted laterally, and with only a very slight transverse median impression .. .. (*inermis*.
{ *aristoteli*.

a a.

Club long, more or less laminate and ensiform.

b b.

Prothorax bipartite, the anterior part much more raised than the posterior.

c.

Posterior tibiæ slender.

(Elytra ferruginous red.)

Club long, not much curved outwardly, apex rounded and slightly narrower than the base; elytra with a black diagonal band, a post-median patch and an apical spot of the same colour *signatipennis*.

(Elytra blue black, edged with ferruginous.)

Club long, broad, slightly curved, outer margin broadly grooved, groove alveolate, upper edge with six serrations, intervals of serration setulose .. .. *concinnus*.

Club a little thicker in the outer than in the inner part, a little curved, not narrower at apex than at base and having a narrow groove in the outer margin reaching from the base to two-thirds of the length .. .. *schaumi.*

Club similar to the preceding one, but a little shorter and a little narrower at apex.. .. .. *germari.*

c c.

Posterior tibiæ broadly dilated.

(Elytra piceous black, edged with ferruginous.)

Club subensiform, very long, outer margin with a moderately wide groove reaching from the base to near the apex, edges of the groove faintly notched .. .. *raffrayia.*

a a a.

Club long, narrow, recurved, outer margin deeply scooped from base to past the median part or near the apex, upper edge subdenticulate.

c.

Posterior tibiæ slender.

(Elytra blue black, edged with ferruginous.)

Club with four impressions on the upper posterior declivity crenating the upper edge of the outer margin, outer basal angle long and sharp .. .. *lineatus.*

Club with six impressions crenating the upper edge of the outer margin, lower edge of posterior groove sinuate, outer basal short, sharp .. .. *afzelii.*

a a a a.

Club scythe-shaped.

Club compressed, outer margin not grooved, basal angle long and sharp; prothorax with a deep, round impression on each side, anterior part not much raised .. .. *cultratus.*

Club rounded, not compressed, more lanceolate than falciform but curved, swollen at base and narrowed in a very sharp point, no outer basal spur; prothorax nearly cleft in two in the median part; elytra with a deep supra-lateral groove on each side .. .. *granulatus.*

a a a a a.

Club straight, narrow, sublaminiform.

c c.

Posterior tibiæ broadly dilated.

Club sharp in the inner part and thicker in the outer, the margin of which is broadly grooved from base to apex with six impressions in the groove, and six denticulations in the upper edge .. .. *klugi.*

a a a a a a.

Club of antennæ broad, dilated, and broadly excavated outwardly.

c c.

Posterior tibiæ broadly dilated.

Head with a raised elevation on the vertex and two small pits, inner margin of club lamelliform and with four striæ, outer margin broadly excavate for nearly all the length and striate on the upper declivity as well as in the excavation, outer basal angle broad; anterior part of prothorax lenticular, incised in the middle, but not aculeate laterally .. .. *cucullatus*.

Head plane and with two small tubercles on vertex; club as in the preceding species but with the apical part a little more acuminate and with the outer basal angle sharper and longer; anterior part of prothorax with a sharp lateral spine .. .. .. *ruber*.

Club compressed and laminate for half the length, the outer margin broadly scooped out at the apex, the upper edge with five slightly projecting ridges; anterior part of prothorax subaculeate laterally .. .. *cochlearius*.

Club with the inner margin sharp, quadri-impressed, a little sinuate, outer margin scooped from base to apex, and with five striæ denting slightly the edges; anterior part of prothorax not aculeate .. .. *viator*.

c.

Posterior tibiæ slender.

Club with two basal striæ in the inner margin, outer margin broad, excavate and with five ridges in the excavation projecting as a rounded denticulation beyond the edges, basal outer angle as long as the whole base, narrow and cylindrical; anterior part of prothorax with a sharp spine on each side .. .. *burchellianus*.

Club laminate in the inner and basal margins, with the outer margin enlarged and broadly excavate from the basal angle to the apex, lower edge of the excavation broader than the upper, the latter sinuate near the apex, both edges slightly striate inwardly; head with four rounded tubercles on vertex; posterior tibiæ a little dilated in the median part .. .. *rugiceps*.

Club curving outwardly, convex in the anterior part, broadly scooped in the posterior one, the lower edge of which is slightly sinuate, while the upper one is briefly subdentate; hind tibiæ slender .. .. *degeeri*.

a a a a a a a.

Antennæ thick, deeply and broadly grooved across the upper part.

c.

Posterior tibiæ not dilated.

Club with two small, closely set sinuations on the inner margin, basal part nearly straight, apical part of upper surface broadly grooved diagonally .. .. *linnei*.

c c.

Posterior tibiæ broadly dilated.

Club with the inner margin slightly bi-impressed, the outer one strongly bisinuate, upper surface with a broad but not deep excavation near the apex and a deep impression about the median part .. .. *burmeisteri*.

a a a a a a a a.

Antennæ long, slender, cylindrical.

c.

Posterior tibiæ slender.

Club a little bent outwardly in the median part, slightly thickened at tip and with two very small teeth at the apex of the outer margin, basal part deeply incised .. .. *curtisi.*

Club very long, straight, slightly thickened at tip, outer basal angle sharp, moderately long .. .. *cylindricornis.*

Club moderately long, apical part of outer margin very slightly grooved, outer basal angle not projecting .. .. *schuckardi.*

Head without a vertical spine.

Pygidium with thick, stiff bristles.

Club massive, broadly excavate in the posterior part, and with a thick tuft of hairs at the apical part of the excavation .. *marshalli.*

P. HUMBOLDTI, Westw.,
Plate XIII., fig. 11.

Trans. Entom. Soc. Lond., 1852, p. 90; Thesaur. Entom. Oxon., 1874, p. 83, pl. xix., fig. 11.

P. ayresi, Pér., Trans. S. Afric. Phil. Soc., vol. iii., 1885, p. 83, pl. i., fig. 5.

Head, antennæ, legs, and prothorax piceous red. elytra and pygidium ferruginous red; head with a very long and sharp conical tubercle; antennal club thick, somewhat oval, carinate all round, inner margin slightly sinuate, outer angle produced in a moderately long, sharp, stout, slightly curving spine; it is thicker in the middle, and has four shallow striæ extending on the upper part of the outer declivity from about the median part towards the apex; the club is smooth in the basal part only, the rest is very briefly pubescent; prothorax divided in two by a broad transverse grooved impression; the anterior part is a little emarginate in the centre and truncate behind, the posterior is as broad as the anterior, but a little less abruptly truncate, and somewhat angulate laterally; elytra subparallel, shining, and with regular series of very short, slightly flavescent, moderately closely set hairs; tibiæ compressed, broad, but not dilated. Length 11–11½ mm.; width 4½–4¾ mm.

Hab. Natal (D'Urban), Transvaal (Rustenburg).

P. DAMARINUS, Westw.,
Thesaur. Entom. Oxon., 1874, p. 84, pl. vii., fig. 9.

Piceous red, with the posterior part of the elytra from the median part castaneous; head with a sharp, long, conical spine; antennal club thick, but a little compressed in the basal part of the inner declivity, marginate all along the inner apical part and slightly sinuate near the apex, slightly incurved near the apical part, which thus appears as if it were a little curved, posterior margin grooved for two-thirds of the length, with the edges dented slightly by the intervals of four shallow striæ occurring on both upper and under sides of the outer declivity; prothorax with the transverse median impression reaching from side to side, and with a lateral flavescent pubescence; anterior part rounded laterally, narrowing from base to apex and slightly emarginate in the centre, posterior part bi-impressed in the centre and constricted laterally at base; elytra parallel, covered with closely set, briefly pubescent punctures; tibiæ slender. Length 10¾ mm.; width 4½ mm.

Hab. Transvaal (Rustenburg), Bechuanaland.

P. MIMUS.

The colour and shape of head and prothorax are as in *P. damarinus*, but the shape of the antennal club is different; it is not so thick in the posterior part, the inner margin is sharper, nearly straight, the emargination of the posterior part is not so pronounced, the grooved part is deeper, and the striæ better defined and indent more the edges of the groove; instead of being smooth and shining they are very finely granulose, opaque, and glabrous; the spur of the outer angle is a little longer and less curved.

The shape of the club is intermediate between that of *P. damarinus* and *P. dohrni*, but it is not setose as in the latter, nor is the posterior margin grooved from base to apex. Length 10½ mm.; width 4 mm.

Hab. Transvaal (Rustenburg).

P. DOHRNI, Westw.,
Trans. Entom. Soc. Lond., 1852, p. 93; Thesaur. Entom. Oxon., p. 92, pl. xvii., fig. 12.

Dark chestnut; elytra covered with a very brief, closely set pubescence; head as in the two preceding species; antennal club longer than broad, thick, setulose, carinate in the anterior and apical margins, the former very slightly sinuate above the base, posterior part broadly grooved from end to end, and with a lateral yellow pubescence; posterior declivity with four broad striæ on both sides, the intervals of which indent the edges of the groove; prothorax shaped as in *P. damarinus* and *P. mimus*, but the anterior part is more regularly rounded laterally.

I have not seen this species, and the diagnosis here given is made from the excellent figure given by Westwood in the "Thesaurus Entomologicus Oxoniensis." The characters distinguishing this species from *P. damarinus* and *P. mimus* are the broader and longer groove in the posterior part of the club, and also the striæ on both sides of the posterior declivity, which are broader and deeper. Length 8 mm.

Hab. Natal, teste Westwood.

P. SPINICOXIS, Westw.,
Proc. Linn. Soc., ii., 1849, p. 59; Thesaur. Entom. Oxon., p. 84, pl. xviii., fig. 7.

Ferruginous red; head with a sharp, conical spine on the vertex; antennal club very briefly pubescent, moderately thick, subelongato-ovate, carinate from the base of the anterior margin to the posterior one, the latter with a faint groove, posterior part of the declivity with four longitudinal, shallow striæ on each side, reaching but not indenting the rounded posterior margin; prothorax transversely impressed in the middle from side to side, and with a lateral yellow pubescence, anterior part rounded, attenuate laterally towards the neck and not emarginate in the centre of the basal part, posterior part deeply scooped in the middle almost up to the base; elytra elongate, nearly parallel, closely punctured, and glabrous; pygidium closely punctured; anterior femora subclavate; intermediate coxæ with a very small spinous process at the base. Length 7-8½ mm.; width 2¼ mm.

Hab. Natal (D'Urban, Maritzburg), Transvaal (Rustenburg), Zambezia (Buluwayo), Mozambique (Rikatla).

P. PROPINQUUS, Pér.,
Plate XII., fig. 7; Plate XIII., fig. 7.
Trans. S. Afric. Phil. Soc., iv., 1886, p. 83.

Chestnut colour, subopaque; head briefly pubescent, and with a sharp, conical spine on the vertex; antennal club subelongato-ovate, moderately thick, and sharply carinate from the angle of the inner margin to the posterior basal angle, which is produced in a short, subtruncate tooth; it is finely granulose, and has a series of very short bristles on each side of the posterior margin; prothorax with a median subdiagonal transverse impression; anterior part convex, a little more raised than the posterior one, rounded laterally, punctulate, and covered with moderately long bristles, posterior part briefly pubescent, a little narrower than the anterior, and with a shallow median depression; elytra a little lighter in colour than the head and prothorax, and covered with a closely set and dense reddish pubescence; legs densely bristly. Length 8-9 mm.; width 3-3¼ mm.

Hab. Transvaal (Bloemhof, Potchefstroom, Heildeberg, Pretoria), Bechuanaland.

P. RUSTICUS, Pér..
Plate XIII., fig. 10.
Trans. S. Afric. Phil. Soc., iii., 1885, p. 82.

Reddish, shining; head with a very conspicuous conical spine on the vertex; antennal club subelongato-ovate, moderately thick, inner and apical margins carinate, the former nearly straight, outer margin slightly emarginate near the apex, also carinate, but with three very small and hardly noticeable dents in the median part of the carina; posterior basal angle not longer than the anterior; prothorax with a very shallow transverse median impression reaching from side to side, and much constricted laterally in the middle, anterior part convex, subcordiform, briefly pubescent, posterior part narrower than the anterior, subcylindrical, and without any median impression; elytra parallel, very finely and closely punctured, each puncture bearing a very short hair; posterior tibiæ slightly dilate. Length 8½ mm.; width 3 mm.

Hab. Transvaal (Rustenburg).

P. ARDUUS, Pér.,
Plate XII., fig. 8; Plate XIII., fig. 6.
Trans. Entom. Soc. Lond., 1896, p. 149.

Red, shining; head with a long conical tubercle on the vertex; smooth, but slightly punctured behind; antennal club short, thick, convex on both sides, carinate all round, depressed at the base with the basal outer angle produced in a long, sharp, slightly recurving spine, no longitudinal impression in the posterior declivity; prothorax smooth, with a deep transverse impression reaching from side to side and having a yellow pubescent patch at each end, anterior and posterior part equally broad, the anterior a little more raised than the posterior, the sides nearly parallel; elytra elongate, subparallel, smooth, and very closely punctured, the punctures in the anterior part being deeper and broader than those behind; tibiæ slender. Length 8 mm.; width 2½ mm.

Hab. Zambezia (Manica).

P. FALLAX, Pér.,
Trans. S. Afric. Phil. Soc., vi., 1892, p. 108.

Head, antennæ, and prothorax chestnut brown; elytra lighter red; head pubescent, and with a short, conical tubercle on the vertex; antennal club shining, slightly pubescent, subovate, carinate all along the inner and apical margins, but emarginate at the apical part of the outer margin, which is grooved, although neither deeply nor broadly, from the emargination to the outer angle, which is produced in a long, triangular spur; prothorax cleft in two by a very deep groove, having a small flavescent patch on each side; anterior part setose, much raised, thick, sloping towards the neck, and constricted laterally; posterior part deeply excavated in the central part, and bituberculate on each side; elytra subparallel, punctured, punctures deep and setigerous; tibiæ slender, arcuate; femora strongly clavate, and very slender at base. Length 5 mm.; width 1¾ mm.

Hab. Transvaal (Potchefstroom).

PAUSSUS BOHEMANI, Westw.,

Trans. Entom. Soc. Lond., 1855, p. 83; Thesaur. Entom. Oxon., p. 93, pl. xviii., fig. 9.

Light brick-red, turning to flavescent in the elytra; head pubescent, and with the whole posterior part raised in a sharp, conical spine; antennal club shining, briefly pubescent, thick, semicircular in the inner part, which is not carinate and ends in a sharp point at the apex, outer margin narrowly but deeply grooved from the apical recurved spine to a short distance from the basal angle, which is produced in a long, slightly recurved spine; prothorax cleft in two by a very deep transverse groove, having a small flavescent patch on each side, anterior part much raised, compressed, thin, nearly perpendicular, carinate at tip, and has a very long pubescence, posterior part deeply excavate in the central part, and bituberculate on each side (Westwood, *loc. cit.*, has given a good side-view figure of the prothorax of this species, but the prothorax of the insect (fig. 9) is not at all correct); elytra punctulate, each puncture bearing a very long hair; pygidium with a long pubescence; tibiæ arcuate, femora strongly clavate, and very slender at base. Length 5¼ mm.; width 2 mm.

Hab. Cape Colony (Kimberley).

PAUSSUS MANICANUS, Pér.,

Plate XII., fig. 2; Plate XIII., fig. 4.

Trans. Entom. Soc. Lond., 1896, p. 149.

Reddish brown, shining; head quite flat and smooth on the vertex; antennal club glabrous, shining, long as the head and prothorax, nearly oblong, a little narrower at apex and base than in the middle, compressed but thick in the median part, carinate all round and with the posterior angle produced in a sharp, moderately recurved spine; prothorax constricted laterally past the median part and with a shallow median transverse impression, having a very small flavescent patch on each side but not incising the lateral parts, the anterior part is depressed, hardly more raised than the posterior, and has a median longitudinal shallow impression, the posterior part is nearly plane, and not impressed in the centre; elytra parallel deeply and closely punctured, each puncture bearing a very short, flavescent seta; femora a little swollen; tibiæ slender. Length $8\frac{1}{2}$–9 mm.; width 3 mm.

Hab. Zambezia (Manica, Buluwayo).

P. INERMIS, Gerstäck,

Monatsb. Berl. Acad., 1855, p. 268; Peter's Reis. n. Mossamb., 1862, p. 268, pl. xv., fig. 12; Westw., Thesaur. Entom. Oxon., p. 95, pl. xix., fig. 5.

Reddish brown, moderately shining; head plane; antennal club subovate, but deeply emarginate in the posterior margin near the apex, carinate all round, not grooved behind, basal part nearly straight, but with a narrow transverse incision in the outer apical angle, the external part of which is produced in a moderately long recurved spine; prothorax elongato-cordate, faintly impressed transversely at about the median part, anterior part convex, posterior part not depressed; elytra subparallel, very closely and finely punctured and briefly pubescent; tibiæ slender. Length 8 mm.

Hab. Mozambique (Tette).

This description is made from the figure in Westwood's 'Thesaurus Entomologicus Oxoniensis.' I am not aware that this *Paussus* has been met with since its capture by Dr. Peters, 1842–1848. The type is in the Berlin Museum.

PAUSSUS ARISTOTELI, Jas. Thoms.,

Archiv. Entom., i., 1856, p. 403, pl. xxi., fig. 2.

"Light chestnut brown; head projecting, strongly and suddenly depressed behind; eyes large, rounded; basal joint of antennæ elongate, second very large, dilate, subconical, or claviform, wider at the base which has a curved spine in the outer angle; prothorax subcordiform, wider in the middle with the anterior angles rounded, divided in the middle by a transverse line, median longitudinal line not much noticeable; elytra at least three times as long as the prothorax and broader at the base, rounded at the humeral angle, nearly truncate at the apex, and with two slight projections on the suture after about the fourth part of the length, dorsal part slightly punctured; legs strong; abdomen punctured, the other parts of the body smooth." Length 8 mm.

This species, which I have not seen, is evidently a close ally to *P. inermis;* but judging from the figure, the club is more compressed, the spur is longer, but apparently also incised, the elytra are more thickly pubescent than in both the figures of *inermis*, and the anterior part of the prothorax is more dilated.

Hab. Natal, teste Thomson, and Port Natal and Abyssinia, teste Raffray.

PAUSSUS SIGNATIPENNIS, Pér.,
Plate XIII., fig. 2.

Trans. S. Afric. Phil. Soc., iii., 1885, p. 83, pl. i., fig. 4.

Brownish red, moderately shining; head, prothorax, elytra, and legs densely pubescent; head hexagonal, plane, a little scooped at apex; penultimate joint thick, flattened, very pilose; antennal club compressed, sublamelliform, as long as the head and anterior part of prothorax put together, nearly as broad as the anterior part of the head, curving outwardly, carinate in the inner margin, slightly emarginate at about the median part in the posterior margin, which is not grooved thickly but briefly pubescent, outer angle produced in a more or less rounded spur; prothorax bipartite, anterior part much raised, almost perpendicular behind, broadest in the middle, sides narrow and ridge shape, posterior part with three longitudinal impressions separated by two median ridges, lateral walls raised and tuberculated at apex above the small flavescent patch; elytra subparallel, somewhat roughly punctured and having a dense, long, greyish pubescence; they are of the same colour as the prothorax and more shining than the head and prothorax, and have on each side a narrow black band running diagonally from under the humeral angle to a short distance of the median part of the suture, a subquadrate patch of the same colour in the posterior part, and the apical part of the suture is also edged with black; tibiæ compressed, slightly dilated. Length 8 mm.; width 2½ mm.

Hab. Transvaal (Potchefstroom).

PAUSSUS CONCINNUS, Pér.,
Plate XII., fig. 6; Plate XIII., fig. 12.

Trans. Entom. Soc. Lond., 1896, p. 150.

Head, prothorax, and legs brick-red; elytra black, edged with red at the base and apex; head and prothorax glabrous, the former edged in front on the vertex with a high semicircular ridge which is broadly emarginate in the middle and reaches from eye to eye, posterior part raised above the neck into a ridge higher than the anterior one, and with a median and two lateral sinuations; median part of head plane, and with a broad depression above each eye partly edged by a very narrow groove which extends also along the posterior raised part; antennal club long, broad, compressed, as long as the base of the prothorax, slightly curving in the inner edge, which is distinctly marginate and has besides a distinct raised line running parallel to it, outer margin broadly grooved from base to apex, and having seven round alveolæ as well as six obtuse serrations on the upper edge, the intervals of which bear each a short yellowish seta, outer basal angle of the club long and sharp; prothorax bipartite, the anterior part much raised, short, abruptly truncate, deeply incised in the centre and on each side, and with a deep transverse impression, the posterior part is narrower than the anterior, and has the shape of a truncate cone, broadly scooped out in the anterior median part, with each side of the incision produced in a sharp, short tubercle; elytra closely set with very short, greyish hairs, but very indistinctly punctured; femora not clavate; tibiæ straight, not thickened. Length 6 mm.; width 2½ mm.

Hab. Zambezia (Salisbury).

PAUSSUS SCHAUMI, Westw.,

Trans. Entom. Soc. Lond., 1852, p. 94; Thesaur. Entom. Oxon., p. 94, pl. xix., fig. 6.

P. noraculatus, Pér., Trans. S. Afric. Phil. Soc., vol. iii., 1885, p. 84, pl. i., fig. 6.

Head, antennæ, prothorax, and legs brownish red; elytra dark blue, with a broad basal band and a narrow apical line brownish red; head with two high longitudinal ridges running from the neck to the apex, finely aciculate, briefly pubescent; antennal club compressed, curving outwardly, distinctly pedunculate at base, as long as the head and anterior part of prothorax put together, carinate in the inner margin, curving outwardly, posterior part a little thicker than the anterior, with the outer margin grooved from one-third of the length to the basal part, the outer angle of which is short and sharp, basal joint elongato-quadrate; prothorax with the anterior part in the shape of a short, broadly truncate cone, scooped on each side, produced in the centre of the transverse impression in two narrow ridges nearly connected across the impression with two similar ones in the posterior part, which divide it into three shallow cavities, the external walls of the posterior part end in a round, blackish tubercle surrounded by the yellow pubescent patch, the surface of the whole prothorax is distinctly though very briefly pubescent; elytra subparallel very finely aciculate and with a few scattered very short hairs; legs moderately slender; pygidium not incised in the posterior margin. Length 8 mm.; width $3\frac{1}{4}$–$3\frac{1}{2}$ mm.

Hab. Cape Colony (Vaal River).

PAUSSUS GERMARI, Westw.,
Plate XIII., fig. 14.

Trans. Entom. Soc. Lond., 1852, p. 94; Thesaur. Entom. Oxon., p. 94, pl. xix., fig. 2.

The only difference between *P. schaumi* and *P. germari* seems to consist in the shape of the antennal club, which is a little shorter, a little narrowed towards the apical part and slightly less curved outwardly owing to the outer margin being slightly straighter, the base of the club is much less conspicuously pedunculate, and the space between the two occipital ridges has a faint triangular impression, a trace of which is found in *P. schaumi*; the sculpture and shape and colour of prothorax and elytra are the same as in *schaumi*. Length $6\frac{1}{2}$ mm.; width $2\frac{1}{2}$ mm.

Hab. Natal, teste Westwood.

This species occurs also in Abyssinia; a specimen from that locality agrees very well with Westwood's figure, except that the clava is entirely reddish brown.

PAUSSUS LINEATUS, Thunb.,
Plate XIII., fig. 5.

Act. Holm., 1781, p. 171, pl. iii., figs. 4–5.

P. parrianus, Westw., Trans. Entom. Soc. Lond., 1847, p. 29, pl. ii., fig. 3; Thesaur. Entom. Oxon., p. 91, pl. xvii., fig. 7.

Dark red, moderately shining; elytra blue black, broadly edged all round with dark red; head flat on the vertex, but with the margins raised all round, slightly aciculate and very briefly pubescent, neck very distinct; antennal club much recurved, as long as the head and the anterior part of the prothorax together anterior margin sharp and carinate, posterior part thicker, outer margin carinate from the apex to one-fourth of the length, deeply scooped from there to the base, the outer angle of which ends in a sharp, long spur, the upper edge of the scooped margin has four striations with raised intervals serrating the edge, the lower edge is not dented, and between the end of the groove and the apex there is a small rounded marginal semicircular projection; the first joint is elongato-quadrate, nearly twice as long as broad, and like the club hardly pubescent; prothorax bipartite, lenticular, with the edges not very sharp, incised in the centre and laterally so as to look quadrituberculate, and raised higher than the posterior, which has two median longitudinal impressions separated by a double ridge, with the lateral walls hardly raised, sloping towards the median transverse impression, which has a small, black tubercle on each side next to the small flavescent patch; elytra subparallel, finely shagreened and very briefly pubescent; legs slender. Length 5-6 mm; width 2-2¼ mm.

Hab. Cape Colony (Cape Town and environs).

PAUSSUS AFZELII,

Trans. Entom. Soc. Lond., 1855, p. 82; Thesaur. Entom. Oxon., p. 96, pl. vii., fig. 6.

P. laetus, Gerstäck, Stett. Zeit., 1867, p. 430.

Very similar to *P. lineatus* in shape, sculpture, and colouring, but the shape of the antennal club and of the prothorax is different; the former has the same shape, but is longer and curves more backward, and the posterior declivity on the upper part has six longer striæ instead of four, with the intervals rounded and denting the upper edge; basal spur sharp but not long; the latter is also bipartite, but the anterior raised part is not so lenticular, it is deeply incised in the middle, but not laterally, and the walls of the posterior part are a little more raised and there is only one broad median impression. Length 6-6½ mm.; width 2¼-2½ mm.

Hab. Transvaal (Rustenburg, Leydenburg); occurs also in Abyssinia.

PAUSSUS CULTRATUS, Westw.,

Plate XIII., fig. 17.

Proc. Linn. Soc., 1849, p. 52; Thesaur. Entom. Oxon., p. 86, pl. xix., fig. 1.

? *P. plinii*, Thoms., Arch. Entom., i., 1857, p. 403, pl. xxi., fig. 3.

Light testaceous, shining; head, thorax, and elytra briefly pubescent; head convex on vertex, elongato-quadrate; first joint of antennæ elongato-quadrate and hollowed in the upper part, club compressed, long, falciform, with both inner and outer margins sharp, the inner one nearly straight for two-thirds of the length and sharply curved, inner margin slightly sinuate past the median part, apical part narrowed and very sharp; prothorax with a deep impression on each side of the median part and with a small pubescent yellow patch on each side, anterior part not more raised than the posterior one, which is slightly narrower; elytra parallel and covered with densely set shallow setigerous punctures; tibiæ slender. Length $4\frac{3}{4}$–5 mm.; width 2 mm.

Raffray is of opinion that *P. cultratus* and *P. plinii* are two different species, which he distinguishes by the shape of the antennal club which is nearly alike, but "in *P. plinii* it is longer, narrower, and more falciform, decreasing gradually from base to tip, and the more regular curve belongs to a circle of a wider diameter, and, therefore, the point is longer, more slender and sharper, while in *P. cultratus* the club looks like an elongate square with the sides nearly parallel as far as the tip, which is suddenly curved to form a shorter and more obtuse point." He adds that he possessed the two species.

It is quite true that the figure given by Westwood of the club of *P. cultratus* is broader in proportion to the length than that of *P. plinii*, and that I have not seen any example as yet absolutely similar, but I think that the latter as differentiated by my excellent friend Raffray is the male, and that Westwood has exaggerated the width of the club in his figure of *P. cultratus*.*

Hab. Natal (D'Urban, Maritzburg, Estcourt, Frere), Transvaal (Potchefstroom, Pretoria).

PAUSSUS GRANULATUS, Westw.,

Proc. Linn. Soc., ii., 1849, p. 58; Thesaur. Entom. Oxon., p. 86, pl. xvii., fig. 5.

Light, testaceous glabrous, moderately shining; head granulose and with two small rounded depressions, one on each side of the ocelli; first joint of antennæ swollen at base, a little attenuate at tip, club rounded, swollen at base and tapering gradually into a sharp point with a seta at tip, falcate, and without any basal outer angle; prothorax bipartite, anterior part nearly perpendicular, narrow, emarginate in the middle, and also, but not so deeply, laterally, median excavation very wide, posterior part excavate and with only the lateral subtuberculate walls left; elytra subelongato-quadrate and with a supra-lateral, deep and broad groove running from the humeral angle to the apex, carinate outwardly, and with a faint silky white tinge, the discoidal part of the elytra is granulose and darker than the general colour, and the sides are not so closely granulose; the inner edge of the intermediate and posterior tibiæ are sinuate inwardly, the latter inflated. Length 4 mm.; width 2 mm.

Hab. Cape Colony (Port Elizabeth, Grahamstown), Transvaal (Pretoria).

* Dr. C. A. Dohrn has expressed also (Zur Literat. d. Pauss. Stett. Ent. 1887, Zeit., p. 317) an opinion similar to mine.

PAUSSUS RAFFRAYI, Pér.,
Plate XII., fig. 1; Plate XIII., fig. 3.
Trans. Entom. Soc. Lond., 1896, p. 150.

Piceous black, with the apical part of the elytra and the tarsi reddish brown; head rugulose, with the anterior part deeply impressed in the centre, and the walls of the impression raised in two short tuberculiform processes, posterior part bi-impressed; club of the antennæ subensiform, very long, compressed, external margin with a moderately wide groove reaching from the base to near the apex, both edges of the groove faintly notched; prothorax bipartite, the anterior part raised, smooth, the median transverse cavity wide and deep and having two yellow pubescent patches in the centre, posterior part depressed and with three tuberculated indentations; elytra subparallel, shining, moderately punctured, each puncture with a very short, greyish hair; anterior and intermediate tibiæ slender, posterior tibiæ broadly dilated and flattened. Length 5 mm.; width 1½ mm.

Hab. Natal (Frere).

PAUSSUS KLUGI, Westw.,
Plate XIII., fig. 15.
Trans. Entom. Soc. Lond., p. 85, pl. ix., fig. 2; Arcana. Entom., vol. ii., p. 183, pl. xci., fig. 4.

Ferruginous or piceous red, moderately shining, very briefly pubescent; head plane on the vertex, with a median longitudinal groove, posterior part with two slight protuberances; club elongate, very little shorter than head and thorax together, laminiform, carinate all along the inner and apical margin, posterior part thicker than the anterior, broadly grooved from base to apex, the groove with seven impressions on the lower part and with six sharp, short, briefly setigerous teeth projecting beyond the edge, upper edge not denticulate; prothorax bipartite, anterior part lenticular, but more convex in front than behind, slightly emarginate in the centre and subangular laterally, median incision very deep and broad, posterior part as broad as the anterior, lateral walls short and sharp, median part with three small impressions divided by two smooth, not much raised tubercles; elytra subparallel and with a very brief and scattered pubescence; posterior tibiæ much dilated, triangular, the anterior and intermediate ones slender. Length 6 mm.; width 2 mm.

Hab. Natal (Maritzburg, Estcourt, Frere), Transvaal (Waterberg). In the examples from Estcourt and Frere the colour is piceous red instead of being ferruginous.

PAUSSUS CUCULLATUS, Westw.,
Proc. Linn. Soc., ii., 1849, p. 59; Thesaur. Entom. Oxon., p. 93,
pl. xviii., fig. 6.

Chestnut red, shining, nearly glabrous; head with an elevation on the vertex, encircled by a sharp ridge, and bearing two very small tubercles; club short, broad, with the anterior part compressed, sharp, and with four deep, narrow striæ, outer margin broadly and deeply excavate for nearly the whole length, and with the internal part of the excavation with six moderately deep transverse striæ, the intervals of which form a blunt serration on both edges; prothorax bipartite, the anterior part in the shape of a thin disk, slightly angular laterally, and not incised in the median part, posterior part with the lateral walls sloping at apex and not much raised, median part moderately excavate and with no longitudinal impressions; elytra subparallel, almost glabrous; femora and tibiæ very much compressed and inflated, the posterior tibiæ more dilated than the others. Length $4\frac{1}{2}$ mm.; width $1\frac{3}{4}$ mm.

Hab. Cape Colony (Uitenhage, Albany, Port Elizabeth), Natal (Estcourt, Maritzburg, Frere).

PAUSSUS RUBER, Thunb.,
Vet. Acad. Handl., 1781, t. 2, p. 170.

Reddish brown, moderately shining; head plane and with two very small tubercles on the posterior part; club somewhat similar in shape to that of *P. cucullatus*, but it is not quite so much dilated, the striæ of the inner margin are not so deep, and the excavated part of the outer margin is not so broad, the striæ in the excavation are deeper, but the intervals, although more convex, hardly dentate the edges; the prothorax bipartite, anterior part lenticular, incised in the middle and distinctly spinose laterally, posterior part bituberculate in the centre, lateral walls sharp and well defined; elytra parallel, nearly glabrous; all the tibiæ are dilated, but the posterior ones are much broader than the others. Length $4\frac{1}{2}$–5 mm.; width 2–$2\frac{1}{4}$ mm.

Hab. Cape Colony (Vaal River, Sterkstroom), Zambezia (Limpopo River).

PAUSSUS COCHLEARIUS, Westw.,
Trans. Entom. Soc. Lond., vol. ii., p. 88, pl. ix., fig. 6; Arcana
Entom., vol. ii., p. 189, pl. xciv., fig. 3.

Chestnut brown, briefly but thickly pubescent, subopaque; head plane in the centre, carinate transversely in the anterior edge and with two diagonal ridges behind diverging from the central part towards the hind part of the eyes; basal joint of antennæ broad, subquadrate, club laminiform for half the length, apical part of the outer margin dilated, scooped out, and having internally five striæ with rounded intervals serrating the lower edge, and also, but in a lesser degree the upper one, outer basal angle moderately long and sharp; prothorax bipartite, the anterior one lenticular and emarginate in the centre and laterally, posterior one depressed in the centre, walls sharp and not tuberculate; elytra subparallel; tibiæ of all legs compressed and broad. Length 5¼ mm.; width 2 mm.

Hab. Transvaal (Potchefstroom), Natal (Estcourt), Cape Colony (Uitenhage).

PAUSSUS VIATOR, Pèr.,

Plate XII., fig. 4; Plate XIII., fig. 19.
Trans. Entom. Soc. Lond., 1896, p. 151.

Piceous black, opaque, with the antennæ and legs very dark red; head with three short impressions in the middle of the vertex, the median one of which is the deepest and is bounded by two short ridges; inner margin of the club sharp, quadri-impressed, a little sinuate at tip, the outer one dilated, broadly scooped out from apex to base with the outer basal angle moderately long and sharp; the cavity of the outer margin has six striæ, the rounded intervals of which serrulate the two edges; prothorax bipartite, with the anterior part lenticular and incised in the centre, posterior part long, lateral walls sloping towards the transverse incision, median part slightly incised longitudinally; elytra parallel, nearly glabrous; tibiæ compressed, broad, the posterior ones much dilated. Length 5 mm.; width 1½ mm.

Hab. Natal (Frere, Estcourt).

PAUSSUS BURCHELLIANUS, Westw.,

Trans. Entom. Soc. Lond., 1869, p. 319; Thesaur. Entom. Oxon., p. 92, pl. xvii. fig. 10.

Chestnut brown, moderately shining; head pubescent, plane in the centre, but raised in a small protuberance behind, apex raised in a transverse ridge; club carinate for half its length in the inner margin and with three marginal impressions, the basal one of which is broader and deeper than the others, outer margin broadly dilate and scooped out, and having five deep striæ extending from one edge to the other with the intervals raised, rounded, serrating both the edges, those on the lower one slightly penicillate at tip, outer basal angle very long and cylindrical; prothorax briefly pubescent, bipartite, anterior part perpendicular with the median part laminiform, subquadrate, broadly emarginate in the centre, lateral part spinose, posterior part depressed in the anterior part with the lateral walls tuberculate; elytra parallel and covered with a long subflavescent pubescence; tibiæ slender. Length 5 mm.; width 2 mm.

Hab. Cape Colony (Albany, Sterkstroom).

PAUSSUS RUGICEPS, Pér.,
Plate XIII., fig. 9.
Trans. S. Afric. Phil. Soc., iv., 1886, p. 82, pl. i., fig. 4.

Elytra piceous red, antennæ, head, prothorax, and legs dark red; head rugose, and with a high prominence divided in two by a deep groove and tuberculose at each end; club slightly pubescent, a little curved, inner margin sharp, not impressed, outer margin dilated, broadly and deeply scooped out from apex to base, outer angle sharp and moderately long, cavity smooth, but with five faint serrations on both edges; prothorax bipartite, anterior part laminiform, subrectangular laterally and with the anterior face sloping, posterior part only a little depressed in the central part, walls not much raised and tuberculate; elytra parallel, very finely aciculate and with regular series of very short, distinct greyish hairs; legs briefly setulose; anterior and intermediate tibiæ moderately slender, posterior ones ampliate, but not broader at the apex than at the base. Length 5 mm.; width 2 mm.

Hab. Transvaal (Rustenburg).

The antennal club is nearly similar to that of *P. Degeeri*.

PAUSSUS DEGEERI, Westw.,
Trans. Entom. Soc. Lond., 1855, p. 82; Thesaur. Entom. Oxon, p. 93, pl. xviii., fig. 12.

Fulvous, elytra finely punctured and covered with very short yellowish setæ; head moderately wide and having two small rounded equidistant tubercles between the eyes; antennal club oblong, curved, anterior margin sharp, rounded at apex, posterior one grooved, the groove elongate, subpyriform, the upper edge with five small rounded tubercles, lower edge a little wider than the upper and simple; prothorax sub-bipartate, anterior part hardly broader than the head, angular, raised, subemarginate in the middle with the outer sides angular, posterior part narrower and with the outer sides raised and parallel, grooved tranversely at about the median part, but not deeply, and having in the middle two contiguous tubercles; elytra much larger than the prothorax, subparallel; legs elongate, slender. Length $6\frac{1}{2}$ mm.

Hab. Caffraria.

I have not met yet with this species, and the description here given is culled from Westwood's.

Paussus burmeisteri, Westw.,
Plate XIII., fig. 16.

Trans. Entom. Soc. Lond., ii., p. 86, pl. ix., fig. 3; Arcana Entom., vol. ii., p. 171, pl. lxxxix., fig. 2.

Opaque and set with squamiform hairs; head, antennæ, prothorax, and legs dark brown; elytra deep chestnut brown; head with an occipital protuberance with a narrow rounded ridge enclosing a small pit; club thick, longer than broad, carinate all round, inner margin slightly bi-impressed, the outer one strongly bisinuate, apical part of the upper surface deeply and broadly scooped transversely near the apical part, the lower edge of the cavity slightly serrulate and very deeply impressed at about the median part, outer basal angle short and sharp; prothorax bipartite, anterior part in the shape of a broadly truncate cone, incised in the middle, posterior part with a very deep median impression and with four small tubercles, two in the middle and one at the apex of each lateral wall; elytra subparallel, set with distant seriated squamiform hairs; all tibiæ dilated, the posterior ones broader than the others. Length 6 mm.; width 2 mm.

Hab. Cape Colony (Cape Town).

Paussus linnei, Westw.,
Plate XIII., fig. 18.

Trans. Linn. Soc., xvi., p. 634, pl. xxxiii., fig. 22; Arcana Entom., ii., p. 169, pl. lxxxviii., fig. 4.

Chestnut brown, head and prothorax a little darker, hind part of the head raised in a small prominence containing two small pits close to one another and with edges carinate; basal joint of antennæ thick, elongato-quadrate; club thick, broad at the base, which is sinuate, with the outer angle sharp but not spinose, inner margin compressed, outer part swollen and broadly and deeply scooped out transversely between the apex and the median part, the excavation is concave and the external wall thin, incurved, and slightly pubescent along the edge; prothorax bipartite, anterior part lenticular, very slightly emarginate in the centre and subsquamose, posterior part deeply excavate in the middle and with two small median and one lateral subtuberculiform processes; elytra subparallel, shining, subsquamiform; legs slightly pubescent; all tibiæ compressed, somewhat broad, but not dilated. Length 4 mm.; width $1\frac{1}{4}$ mm.

Hab. Cape Colony (Cape Town, Oudtshoorn).

PAUSSUS BARKERI, Pér.,
Plate XII., fig. 5; Plate XIII., fig. 13.
Trans. Entom. Soc. Lond., 1896, p. 152.

Reddish brown with very short pubescence all over; head with two median carinæ aculeate in front and overlapping the point of insertion of antennæ, these two ridges diverge slightly from the middle of the vertex and have a narrow groove ceasing abruptly above the neck, which is very short and not constricted; basal joint of antennæ quadrate, very thick; club moderately long, not compressed, deeply sinuate in the inner part, which, like the rounded apical part, is acutely marginate, outer part also sinuate and having on the margin four very distinct teeth bearing several very short setæ, while the apical angle is developed into a long, broad, blunt spur, the inner part of which curves so as to form a short tooth corresponding to a similar tooth situated on the opposite part of the base; the joints of the antennæ are covered with closely set, very short, squamiform hairs; prothorax bipartite, the anterior part ridged, slightly grooved in the centre, posterior part as broad as the anterior with a broad median depression nearly reaching the base; elytra subparallel, covered with very short, closely set hairs, apparently thicker than those on the prothorax and without punctures; pygidium thickly pubescent and with a fringe of long, thickly set, yellowish hairs; anterior and intermediate femora and tibiæ slender, posterior femora and tibiæ dilated and compressed. Length 9 mm.; width 3½ mm.

Hab. Natal (D'Urban).

PAUSSUS CURTISI, Westw.,
Proc. Entom. Soc. Lond., 1864, p. 190; Thesaur. Entom. Oxon., p. 84, pl. xviii., fig. 11; Raffray, Matér. Etude Pauss., p. 32 pl. viii., figs. 35, 36.

Chestnut brown, subopaque, glabrous; head with two sharp, median ridges in the anterior part uniting in the centre with the apex of a bisinuate, subtriangular one which reaches from side to side in the posterior part, and another but shorter one in the base adjoining the neck, there is also a lateral one running above the eye, these ridges enclose thus three deep impressions in the anterior part and two smaller ones in the posterior; the genæ are distinctly aculeate; antennæ densely squamiform, basal joint elongato-quadrate club long, slender, cylindrical, curving, slightly thickened at the tip, which is carinate, outer margin with a very short groove and two small teeth at the apex only, base deeply incised, outer angle blunt; prothorax bipartite, anterior part in the shape of a broadly truncate

cone, broadly emarginate in the middle, posterior part divided by a
deep, narrow, transverse groove, deeply impressed up to the base,
lateral walls not tuberculate; elytra elongate, glabrous; tibiæ very
little compressed, nearly cylindrical; legs densely squamiform.
Length 8 mm.; width 2½ mm.

Hab. Transvaal (Potchefstroom), Natal (Estcourt), Cape Colony
(Port Elizabeth); occurs also in Abyssinia.

PAUSSUS CYLINDRICORNIS, Pér.,
Plate XIII., fig. 1.

Trans. S. Afric. Phil. Soc., iii., 1885, p. 81, pl. i., fig. 2.

Reddish brown, subopaque; head slightly squamose with two
median ridges reaching from the apex to the posterior part, which
is slightly raised and has two contiguous pits on each side of the
head, and running above the eye is a shorter ridge running from the
neck to some distance from the apex; eyes prominent and not
bordered by the genæ; antennæ densely squamose, basal joint short,
quadrate, club long, nearly cylindrical, a little compressed at the tip,
which is very slightly ampliate and carinate in the rounded part,
base not incised, outer angle moderately long and sharp; prothorax
bipartite, anterior part in the shape of a broadly truncate cone,
widely emarginate in the middle and shorter than the posterior one,
which is broadly grooved longitudinally in the centre, both parts are
slightly squamose; elytra elongate, parallel and covered with densely
set, squamiform, subflavescent hairs; legs bristly; tibiæ linear.
Length 8½ mm.; width 2½ mm.

Hab. Transvaal (Rustenburg), Bechuanaland.

PAUSSUS SCHUCKARDI, Westw.,

Proc. Entom. Soc. Lond., ii., p. 87, pl. ix., fig. 4; Arcana Entom.,
vol. ii., p. 187, pl. xcii., fig. 5; Raffray, Matér. Etud. Pauss.,
p. 32, pl. viii., figs. 30, 31.

Reddish brown, subopaque; head grooved in the central parts with
the edges of the groove rounded, raised and reaching the posterior
part, above the eye there is also a small ridge on each side, and the
space between this supra-ocular ridge and the median one is depressed, genæ projecting a little; antennæ densely squamiform,
basal joint thick, elongato-quadrate, club quite cylindrical, not quite
truncate at tip, the apical margin carinulate; prothorax deeply impressed tranversely in the middle but not exactly bipartite, anterior
part nearly rounded but still slightly more raised than the posterior,
which has a moderately broad, not very deep longitudinal impression;
elytra elongate, subparallel with squamiform hairs not densely set;

legs very briefly pubescent; tibiæ linear. Length 6¾-7 mm.; width 2½-2¾ mm.

Hab. Cape Colony (Grahamstown, Vaal River, Queenstown), Transvaal (Bloemhof, Rustenburg).

In the female the antennæ are a little shorter than in the male.

PAUSSUS MARSHALLI, Pér.,
Plate XII., fig. 3; Plate XIII., fig. 11.
Trans. Entom. Soc. Lond., 1896, p. 153.

Reddish brown, shining, elytra thickly pubescent; vertex of the head nearly plane, posterior part ridged above the neck and along the outer sides; basal joint of antennæ quadrate, nearly as large as the head, club broad, massive, inner margin carinate with a short, round basal spur not projecting much, outer margin broadly hollowed with the edges bisinuate, acute at the apical part of the cavity, and bearing on each side a dense tuft of long, yellowish hairs, basal outer spur very broad and subquadrate; prothorax bipartite, the anterior part compressed in a sharp ridge, slightly emarginate in the centre and subaculeate laterally, the posterior part hollowed anteriorly and with a triangular longitudinal groove, and the outer sides produced in a carina sinuate in the middle, sharp in the anterior part, and with the posterior part forming a long tooth standing at an angle with the base, both the points of the lateral carina having a distinct tuft of hairs; elytra short, subparallel; pygidium with short pubescence and having in the middle three transverse rows of long and very thick bristles; legs slender, anterior femora not thickened. Length 5½ mm.; width 2½ mm.

Hab. Natal (Frere, Estcourt).

Mr. A. E. Haviland writes that he found this species drowned in a trough in company with two kinds of ants.

PAUSSUS BRAUNSI.

Chestnut brown, moderately shining; head hardly pubescent, not depressed in the centre, and having two small ocelli-like cavities in the posterior part; antennæ very slightly and very briefly pubescent, inner margin of the club sinuate in the middle, sharply carinate, and without any transverse impression, outer margin broadly dilate, and scooped out from apex to base, cavity with four striæ, the intervals of which are raised but hardly indenting the edges, basal angle produced in a long spur, subquadrate at base and cylindrical from the middle to the apex; prothorax bipartite, and similar in shape to that of *P. burchellianus* and *P. cucullatus;* posterior part of the disk briefly pubescent; elytra parallel, shining, and having regular series of very short, pallid hairs, outer and posterior margins fringed with a series of long, thick bristles, curving backwards on the outer margin; pygidium with a few short bristles on the lower edge; legs slender, briefly pubescent, anterior and intermediate femora very slender at base, but not clavate, almost cylindrical like the tibiæ, posterior ones compressed and only slightly dilated, posterior tibiæ also slightly dilated and compressed. Length 3½ mm.; width 1¼ mm.

In shape the posterior margin of the club is nearly similar to that of *P. burchellianus* and *P. cucullatus*, but it is less broadly scooped than in both these species, and the basal spur is shorter than in the first-named species and not entirely cylindrical, the shape of the anterior margin is however very different, being sublinear owing to a median sinuation, and not impressed transversely; the disposition along the outer and posterior margins of thick, recurved, stiff bristles is unique among the South African *Paussi*, of which it is also the smallest.

Discovered in the nest of *Pheidole capensis* by Dr. Brauns at Port Elizabeth, Cape Colony.

GEN. HYLOTORUS, Dalman,
Analect. Entom., 1823, p. 103.

Head round, with two large fossæ on the vertex excavated in front for the reception of the antennæ, the latter are two-jointed, the first joint is very small, the other lanceolate; maxillary palpi four-jointed, second joint very broad, rounded outwardly, sinuate inwardly, maxillæ without outer lobe, bifid at tip; last joint of maxillary palpi long, attenuate and rounded at tip; prothorax subcylindrical, attenuate behind; elytra subparallel; first abdominal segment very wide, second and third very small, third wider than the preceding and emarginate; legs short, broad, compressed, the femora grooved so as to allow the insertion of the laminated inner part of the tibiæ when retracted; tarsi short, equal.

Two species other than the South African one are known—one from Sierra Leone, and the other from Abyssinia.

HYLOTORUS HOTTENTOTUS, Westw.,
Thesaur. Entom. Oxon., p. 81, pl. xvii., fig. 1.

Chestnut red, glabrous, moderately shining; head rugulose and with two impressions on the vertex containing a round tubercle perforated at tip; eyes not projecting, reniform; no neck; prothorax cylindrical, posterior part narrowed from the middle to the base, discoidal part with a linear, transverse, shallow impression; elytra a little rounded at the shoulder, straight laterally and a little ampliate from the shoulder to the apex, smooth and glabrous. Length 6 mm.; width 3 mm.

Hab. Natal (Maritzburg).

POSTCRIPT.

[The species described below belongs to the genus *Paussus*, Linn. See page 15.]

PAUSSUS ELIZABETHÆ.

Ferruginous red, moderately shining; head foveate, the foveæ shallow and bearing a short greyish seta, vertex with a slightly conical protuberance in the posterior part and two parallel elongate, ocelli-like, deep pits, with smooth raised edges in front of the conical protuberance, and in the centre of the vertex; basal joint of antennæ very rugose and bristly, club swollen, anterior margin compressed, narrow, the anterior edge with a series of yellowish distant setæ on each side, upper part with three deep transverse impressions situated between the base and the median part, posterior margin broadly dilated and deeply scooped, with the basal angle sharp but moderately long, both the upper and lower edges of excavation are symmetrical, but the upper edge has six serrations and the lower one five only, and less conspicuous, the concave part has six broad grooves on each side; thorax rugose, setulose, anterior part in the shape of a disk, very narrowly incised in the centre, and angular laterally, posterior part with the lateral walls sloping, and the median part raised in the centre and grooved longitudinally; elytra rugulose, and with closely set series of moderately long flavescent hairs, outer margins of pygidium clothed with long decumbent pale flavescent hairs; tibiæ of all legs compressed and dilated. Length $4\frac{1}{4}$ mm.; width $1\frac{1}{2}$ mm.

Allied to *P. cucullatus*, and distinguished at once by the shape of the ocelli and the subconical protuberance on the vertex of the head; the antennal club is of nearly the same shape, but not so broad outwardly, and the anterior margin is narrower, and has three transverse impressions instead of four, and the edge of the lower margin of the excavation does not project beyond the upper edge as it does in *P. cucullatus*; it differs also by the setulose elytra, and the fringe of decumbent hairs on the outer margins of the pygidium.

Hab. Cape Colony (Port Elizabeth). Captured by Dr. Brauns.

INDEX TO FAMILY PAUSSIDÆ.

Arthropterus	8, 10
kirbyi	11
Cerapterus	8
smithi	9
concolor	9
laceratus	10
Hylotorus	8, 40
hottentotus	40
Paussus	8, 15
afzelii	19, 30
arduus	18, 24
aristoteli	18, 26
barkeri	37
bohemani	18, 25
braunsi	39
burchellianus	20, 34
burmeisteri	20, 36
cochlearius	20, 33
concinnus	18, 27
cucullatus	20, 33
cultratus	19, 30
curtisi	21, 37
cylindricornis	21, 38
damarinus	17, 21
degeeri	20, 35
dohrni	17, 22
fallax	18, 24
germari	19, 29
granulatus	19, 31
humboldti	17, 21
inermis	18, 26
klugi	19, 32
lineatus	19, 29
linnei	20, 36
manicanus	18, 25
marshalli	21, 39
mimus	17, 22
propinquus	17, 23
raffrayi	19, 32
ruber	20, 33
rugiceps	20, 35
rusticus	17, 24
schaumi	19, 28
schuckardi	21, 38
signatipennis	18, 27
spinicoxis	17, 23
viator	20, 34
Pentaplatarthrus	8, 13
paussoides	14
natalensis	14
Pleuropterus	8, 11
alternans	12
hastatus	12, 13

BIBLIOGRAPHY TO FAMILY PAUSSIDÆ.*

DALMAN, J. W.
    Analecta Entomologica. Holmiæ, 4to, 1823.

DOHRN, DR. C. A.
    Zur Literatur der Paussiden. *Entom. Zeit. Stett.*, vol. xlviii., 1887, pp. 316–318. *Loc. cit.*, 1890; *loc. cit.*, 1891, p. 388.

GYLLENHAL, L.
    Schönherr's Synonymia Insectorum. Appendix, 2 vols., Skara, 1817.

LINNÉ, C. VON
    Bigæ Insectorum (Diopsis, Paussus). *Resp. Andr. Dahl. Decbr.* 1775. Upsaliæ, Edman. 4, pp. 7, c. tab.

PÉRINGUEY, L.
    Notes on three Paussi. *Trans. Entom. Soc. Lond.*, 1883, pp. 133–138.
    Notes on some Coleopterous Insects of the Family Paussidæ. *Proceedings Entom. Soc. Lond.*, 1886, pp. xxxiv.–xxxvii.
    * First Contribution to the South African Coleopterous Fauna. *Trans. S. Afric. Philos. Soc.*, vol. iii., 1885.
    * Second Contribution to the South African Coleopterous Fauna. *Loc. cit*, vol. iv., 1888.
    * Fourth Contribution to the South African Coleopterous Fauna. *Loc. cit.*, vol. vi., 1892, p. 108.

RAFFRAY, A.
    Matériaux pour servir à l'étude des Coléoptères de la famille des Paussides. *Extr. d. Archiv. d. Museum*, 1887, pp. 308–359 and 1–52.

THOMSON, JAS.
    Catalogue des Paussides de la Collection de Mr. James Thomson. *Musée Scientifique*, pp. 67–72. Paris, 8vo, 1860.
    * Archives Entomologiques, 2 vols., 1858. Paris, 8vo.

THUNBERG, C. P.
    Beskrifning pa tvänne nya insecter (Paussus ruber, lineatus). *Vetensk. Acad. Handl.*, 1775, vol. xxxvi., pp. 254–260; ibid., 1781, t. 2, pp. 168–171.

WESTWOOD, O.
    Description of some new or but imperfectly known species belonging to the Coleopterous Family Paussidæ. *Trans. Entom. Soc. Lond.*, vol. ii., 1837–1840, pp. 84–98.
    Arcana Entomologica, vol. ii. London, 1843–1845.
    Description of a new species of Paussus from Southern Africa. *Trans. Entom. Soc. Lond.*, 1847, pp. 29–32.
    Description of seventeen new species of the Family Paussidæ. *Proceedings Linnean Society*, ii., 1849, pp. 55–60, 1850; ii., pp. 100–101.
    Description of three new species of Paussus. *Proceedings Entom. Soc. Lond.*, 1864, pp. 189–190.
    Description of new Exotic Coleoptera. *Trans. Entom. Soc. Lond.*, 1869, pp. 315–320.
    Thesaurus Entomologicus Oxoniensis. Oxford, 1874, 4to, 205 pp., 40 pl.

---

* The works already quoted in the Bibliography of the other families are prefaced by an asterisk.

DESCRIPTIVE CATALOGUE OF THE COLEOPTERA OF SOUTH AFRICA.—Part IV.

By A. Raffray, Memb. Ent. Soc. of France, &c.

Family PSELAPHIDÆ.

The *Pselaphidæ* have in South Africa a very singular faunistic distribution. They appear to be divided in two faunas : one which I call the Cape fauna proper, the other is a general African one : the first seems to be merely restricted to the Cape Peninsula, the second one extends from Mashunaland to Natal.

Unfortunately many parts of South Africa may be considered as unexplored so far as the collection of these minute insects is concerned, and the neighbourhood of Cape Town is the only place where the *Pselaphidæ* have been thoroughly searched for by Mr. Péringuey and myself. They have also been collected in Mashunaland by Mr. G. A. K. Marshall, in Natal by Messrs. Marshall and A. E. Haviland, and in Bechuanaland and the Transvaal by Mons. E. Simon.

Thirty-one genera and seventy-six species are now known to occur in South Africa, and of these fifteen genera and sixty-nine species are not met with anywhere else, so far as our knowledge of these insects goes.

The distribution of these 31 genera and 76 species in South Africa and other parts of the world may prove of interest.

6 genera are found in every part of the world, Europe included, *i.e.* :—

1 *Bryaxis*. It is very doubtful if this genus occurs in South Africa ; the locality of the unique specimen recorded is unknown.

3 *Euplectus, Ryxabis*, and *Pselaphus* occur in Mashunaland and Natal, but not at the Cape.

1 *Reichembachia* is found in Mashunaland, Natal, and the Cape.

1 *Ctenistes* occurs in Natal and at the Cape.

1 genus is found in every part of the world, Europe excepted,
i.e. :—
Tmesiphorus at Natal.

1 genus is found in Northern and Eastern Africa, i.e. :—
Marellus; it has been met also in Natal.

1 genus occurring in Eastern and Western Africa, i.e. :—
Ognocerus occurs also in the Transvaal.

3 genera, *Zethopsus, Batrisodes*, and *Odontalgus*, represented in East Africa and Asia, are found respectively in Mashunaland and Natal, in Natal only, and in Natal and at the Cape.

3 genera, *Asymoplectus, Raffrayia*, and *Trabisus*, which are East African, are represented, the first in Mashunaland and the Cape Colony, the second at the Cape and in Natal, and the locality of the third, although not recorded exactly, is a South African one.

1 genus, *Syrbatus*, a distinctly American one, is represented in Mashunaland.

15 genera are found exclusively in South Africa, of which—

1, *Batoxyla*, is peculiar to Mashunaland.
1, *Novoclaviger*, is peculiar to Mozambique.
2, *Fustigeropsis* and *Commatocerodes*, are peculiar to the Transvaal.
1, *Dalmina*, is peculiar to the Cape and Natal.
10, *Faronidius, Prodalma, Trimyodites, Anoplectus, Xenogyna, Pselaphocerus, Pselaphischnus, Laphidioderus, Pseudotyrus, Fustigerodes*, are peculiar to the Cape Colony.

Out of 16 genera occurring at the Cape (*Reichenbachia, Ctenistes, Odontalgus, Asymoplectus, Raffrayia, Dalmina, Faronidius, Prodalma, Trimyodites, Anoplectus, Xenogyna, Pselaphischnus, Laphidioderus, Pselaphocerus, Pseudotyrus, Fustigerodes*), and of which 10 are peculiar to the Cape, 2 are found in every part of the world, 3 are found in other parts of Africa, and 1 occurs also in Natal.

It is worthy of note that the genus *Raffrayia*, which seems to be a very distinct feature of the Pselaphid fauna of the Cape, where it is represented by 17 species, has only one representative in Abyssinia and another one in Natal; and the genus *Odontalgus* is represented at the Cape by a very aberrant form.

Out of the 12 genera occurring in Natal (*Euplectus, Reichenbachia, Ctenistes, Tmesiphorus, Marellus, Zethopsus, Batrisodes, Odontalgus, Raffrayia, Dalmina, Novoclaviger, Pselaphus*), 5 occur in every part of the world, 3 in different parts of Africa, 2 in other parts of Africa and Asia, 1 in the Cape, and 1 only is peculiar.

Out of the 6 genera occurring in Mashunaland (*Euplectus, Reichembachia, Zethopsus, Asymoplectus, Syrbatus, Batoxyla*), 2 occur in every part of the world, 1 in other parts of Africa and Asia, 1 in America, 1 at the Cape, and 1 is not recorded from anywhere else.

Thus out of 16 genera in the Cape Colony, 10 are special; out of 12 genera in Natal, 1 is special; out of 6 genera in Mashunaland, 1 is special; and 5 of the 12 genera occurring in Natal are also found at the Cape; 2 out of the 5 found in Mashunaland occur also at the Cape.

If we now examine the species, we notice that the isolation of the Cape fauna from that of other parts is still more clearly marked.

Out of the 76 species known to occur in South Africa, 7 of which only are found in other parts of Africa, 2 are from Abyssinia, 2 from the West Coast of Africa, and 3 from Zanzibar. Of the 2 Abyssinian species, 1, *Reichembachia circumflexa*, occurs also in Mashunaland and Natal; the other, *Odontalgus vespertinus*, is also met with in Natal. Out of the 2 species from the West Coast, 1, *Reichembachia picticornis*, is met in Mashunaland and Natal, and the other, *Ctenistes imitator*, in Natal; the 3 Zanzibar species, *Zethopsus sulcicollis, Tmesiphorus rugicollis, Pselaphus longiceps*, are also met with in Natal.

The Cape Colony species number 45, all special; Natal 14, 7 of which are peculiar to that country; and Mashunaland 14, 12 of which are not as yet recorded from elsewhere.

Shape of body variable; elytra short; abdomen free and consisting of six tough segments; maxillary palpi oftener big and quadriarticulate, occasionally inconspicuous and with only one joint, but always provided with a small apical appendage; labial palpi small, biarticulate; coxæ and trochanters variable in shape; tarsi always trijointed, with the first joint extremely small, and having one or two claws either equal or unequal.

The family *Pselaphidæ* is very closely allied to the *Staphylinidæ*, and is seemingly a degenerate form of the latter. It is, however, differentiated from the *Staphylinidæ* by the following characters: the abdomen consists of six segments (except in the male of some species, which have a seventh segment), fused together and therefore immovable; the labial palpi have never more than two joints, and the last joint is provided with at least one small appendage; the last joint of the maxillary palpi is always provided with an apical appendage.

So far as is now known few species are recorded from South Africa. This scarcity is caused by the dryness of the climate. In tropical climes, where damp heat and moisture prevail, the *Pselaphidæ* are very numerous. But although not numerous, they are peculiar to that part of the world, and the South African fauna is a very isolated one, except for the eastern part, where an affinity with the Zanzibar fauna does undoubtedly exist.

Pselaphidæ are found under stones, more especially where the ground is clayey and schistaceous, under the bark of trees, dead leaves, those of oaks especially, in moss, flying at sunset in marshy places, and in ants' nests.

The family is divided into two sub-families, PSELAPHIDÆ GENUINÆ and CLAVIGERIDÆ, divided by the following distinctive characters:—

Maxillæ and paraglossæ spiculose; elytra simple at apex	PSELAPHIDÆ.
Maxillæ and paraglossæ with a long pubescence; elytra and abdomen plicate at base and fasciculate..	CLAVIGERIDÆ.

PSELAPHIDÆ.

All the trochanters short, insertion of the femur on the trochanter lateral, base of femur touching, or nearly so, the coxæ	PSELAPHIDÆ BRACHYSCELIDÆ.
Median trochanters (sometimes the anterior and posterior ones also) long, insertion of the femur on the trochanter apical, and consequently always at a great distance from the coxæ	PSELAPHIDÆ MACROSCELIDÆ.

Synopsis of Tribes.

PSELAPHIDÆ BRACHYSCELIDÆ.

A 2. Hind coxæ prominent and conical; body more or less elongate and depressed.
   B 2. Middle coxæ prominent and conical; tarsi with two claws of generally the same size .. .. FARONINI.
   B 1. Middle coxæ globular, not prominent.
     C 2. Tarsi with a single claw .. .. EUPLECTINI.
     C 1. Tarsi with two claws of very unequal size, the internal one very small, and sometimes hardly apparent .. .. TRICHONYNI.
A 1. Hind and middle coxæ globular, not prominent, the hind coxæ sometimes a little triangular and depressed.
   B 2. First ventral segment very short, always more or less hidden under the hind coxæ or the metasternum.
     C 2. Tarsi with two very unequal claws; hind coxæ somewhat triangular and depressed (but neither prominent nor conical) .. .. BATRISINI.

C 1. Tarsi with a single claw; hind coxæ decidedly globular .. BRYAXINI.
B 1. First ventral segment large, always longer than the hind
coxæ; antennæ geniculate, the first joint very long .. .. GONIACERINI.

PSELAPHIDÆ MACROSCELIDÆ.

A 2. First ventral segment large, longer than the hind coxæ;
tarsi with a single claw .. .. PSELAPHINI.
A 1. First ventral segment short, more or less hidden under the
hind coxæ or the metasternum.
B 2. Epistoma more or less notched, prominent laterally; pubescence short and squamose .. .. CTENISTINI.
B 1. Epistoma simple and not prominent laterally; pubescence
generally long and hairlike, and when short never
squamose .. .. TYRINI.

So far as is now known the following tribes have no representatives in South Africa: *Bythinini*, *Cyathigerini*, *Hybocephalini*, *Schistodactylini*, *Arhytodini*.

TRIBE FARONINI.

Raffray, Rev. Entom., 1890, pp. 82 and 84.

Body linear and depressed; antennæ hardly clubbed at tip; middle and hind coxæ conical and prominent; first ventral segment conspicuous; tarsi with two equal claws.

In general facies the insects included in this tribe are very much like some *Staphylinidæ* of the group *Homalini*. They are somewhat rare, but numerous in New Zealand; they have a few representatives in the South of Europe, in Algeria, and in North America, and only one species is known from South Africa.

GEN. FARONIDIUS, Casey,

Trans. Entom. Soc. Lond., 1887, p. 381; Raffray, Rev. Entom., 1893, p. 3.

Elongate, somewhat depressed; head transverse, without temporal prominences; antennal tubercle large; eyes large, set backward; maxillary palpi middling, second joint clavate, third a little shorter than the preceding one, fourth more than twice as large, ovate, not particularly acuminate at tip, very briefly pilose and with a hardly noticeable apical appendage; antennæ moderately long, moniliform, hardly thickened at tip; prothorax subhexagonal, more attenuate in front than behind, and impressed; elytra much longer than the prothorax and depressed; abdomen strongly marginate and having on the dorsal side five clearly defined segments and on the under side six in the female and seven in the male; metasternum large, subquadrate and convex; legs hardly long; intermediate and posterior coxæ approximate; both claws of tarsi of equal length.

The genus is allied to both *Sagola* and *Faronus*, and differs mostly from them by the eyes, which are very large and situated near the posterior angle of the head. It includes only one species.

Faronidius africanus, Casey,
Trans. Entom. Soc., Lond., 1887, p. 382, c fig. (male); Raffray.
Rev. Entom., 1893, p. 4, pl. 1, fig. 15 (female).

Moderately elongate, depressed, testaceous red, with the antennæ, palpi, and legs testaceous, covered with a moderately dense flavous pubescence; head much transverse and without any posterior angles; eyes large; antennal tubercle prominent, narrow, depressed, and slightly canaliculate; vertex little raised transversely; antennæ half the length of the body, not distant from one another at base, stout, first joint elongate, subcylindrical, second ovate, third smaller, subquadrate, fourth to eighth briefly oblong and slightly decreasing in length, ninth to tenth subquadrate, eleventh oblong and acuminate at tip; prothorax as long as broad, a little wider than the head and eyes, much attenuate in front, rounded laterally in the median part and sinuate from there towards the base, lateral foveæ large, the median one small, and joined by a strong, transverse and arcuate sulcus to two minute, oblong, basal foveæ; elytra more than twice the length of the prothorax and a little longer, sutural stria pluripunctate at base, dorsal stria extending as far as the median part and pluripunctate at base, between the striæ are four punctures disposed in a line; abdomen nearly equal in length to the elytra, the three basal segments gradually increasing in length. The male is distinct from the female; the antennæ are half the length of the body, the joints from the fourth to the tenth inclusive are slightly decreasing in length, and from the fifth to the ninth, a little angular internally, and obliquely subemarginate externally at apex; seventh ventral segment small and with a quadrate median impression. Length 1·30–1·60 mm.

This insect is rather variable, especially in size; it is in the large-size males that the intermediate joints of the antennæ are more or less angular internally; in the females of small size the joints are slightly thicker towards the tip. The median impression of the prothorax is somewhat quadrate with the lateral offshoots very short.

Hab. Cape Colony (Wellington, Stellenbosch, Newlands, Cape Town).

Tribe EUPLECTINI.

Raffray, Rev. d'Entom., 1890, pp. 82 and 91.

Body elongate, more or less depressed; antennæ distant at base or approximate; maxillary palpi variable, sometimes recumbent in an upper fovea; prothorax more or less cordate; elytra variable; abdomen with five conspicuous segments on the upper part and six underneath, sometimes even seven in the male; intermediate coxæ globose, not prominent; posterior ones conical; first ventral segment more or less conspicuous; tarsi triarticulate, with the basal joint very small; one single claw.

This tribe includes a large number of minute insects found in marshy places throughout the world.

The characters of the genera are subject to a great deal of modification, and therefore very inconsistent.

Gen. ZETHOPSUS, Reitter.

Ent. Monatsbl. 1880, p. 85; Raffr., Rev. Entom., 1887, p. 50.

Zethus, Schauf. Nunq. Otios. 11, p. 249.—*nom prevec.*

Linear, depressed; head transverse behind, with a large antennal tubercle in front and a large, superior fovea for the reception of the palpi on each side; eyes placed laterally and backward; maxillary palpi quadriarticulate, first joint inconspicuous, second large, slender at base and strongly clavate, third inserted on the side of the second, smaller and irregular, fourth transverse, irregularly ovate and inserted in the middle of the preceding one; antennæ ten-jointed, first joint large, second much bigger than the following, third to ninth inclusive moniliform, transverse, tenth largest of all, globose or briefly ovate; prothorax more or less cordate, impressed; elytra subquadrate; abdomen elongate, marginate; legs short; intermediate coxæ close to the posterior ones; first joint of the tarsi subconical, second hardly noticeable, third large and with a single claw.

This genus is a most peculiar one; the joints of the palpi fold on each other into a deep fovea situated on the upper surface of the head on each side of the frontal tubercle; the abdominal sexual indices are wanting, but sometimes the frontal tubercles of the male show some difference from those of the female.

It is more largely represented in the Indo-Malayan region than in Africa. I have, however, captured several species in Zanzibar, and one occurs in Mashunaland.

ZETHOPSUS LATICEPS.

Ferruginous, entirely smooth and shining, and covered with a brief, pallid pubescence; head large, transverse, with the antennal tubercle three times as wide as the vertex, slightly rounded at apex and impressed in the middle; first joint of antennæ simple, second quadrate, third to ninth inclusive smaller, transverse and compact, tenth largest of all, subglobose, hardly acuminate; prothorax narrower than the head, cordate, with a large double median transverse fovea, lateral foveæ transverse, the discoidal fovea well distinct and rounded; elytra subquadrate and convex, bifoveate at base and with the external fovea sulciform; abdomen longer than the elytra and a little narrower, first three dorsal segments broadly and deeply impressed at base with the impression filled with an ochreous squamose pubescence; fourth joint a little larger than the preceding ones. Length 1·10 mm.

Owing to the very large frontal tubercle, this species is closely allied to Z. latifrons, Raffr., from Zanzibar, but in this last-named species the prothorax is punctured and the antennæ are longer and the two first joints as well as the last one longer and broader.

Hab. Zambezia (Salisbury).

ZETHOPSUS SULCICOLLIS.

Chestnut brown, with a slight flavous pubescence; antennæ and legs red; head and prothorax with scattered punctures; head not much transverse; antennal tubercle broad, deeply sulcate, dentate inwardly and faintly sulcate on the vertex; the two basal joints of the antennæ quadrate, the first one larger than the others, the third to the ninth inclusive moniliform, transverse, the tenth large, ovate; prothorax subcordate, longer than broad, and having a wide, geminate, median fovea, somewhat shallow, the sides are robust and and there is a deep discoidal sulcus in the anterior part; elytra hardly longer than broad, wider than the prothorax, slightly convex, rounded laterally, and having at the base two large oblong foveæ; sutural stria entire, dorsal one wanting; first three dorsal segments broadly impressed; tibiæ, especially the intermediate ones, slightly thickened in the middle outwardly and arcuate inwardly; female unknown. Length 1·20 mm.

Hab. Natal, also Zanzibar.

Z. sulcicollis differs from the preceding one by the less broad frontal tubercle and the deep groove in the anterior part of the disk.

Gen. PRODALMA.

Body oblong, subconvex ; head large ; antennæ distant from one another, eleven-jointed, club little swollen, last joint larger than the others ; palpi moderately large, first joint hardly noticeable, second slender and clavate at tip, third ovate, minute, fourth large, slightly securiform and acuminate at tip; prothorax cordate, sulcate transversely and bifoveate ; elytra short, attenuate at shoulders, but dentate, lateral margin sulcate, sutural stria entire, dorsal one abbreviate ; first dorsal segment of abdomen larger than the others, female with six ventral ones, male with seven, seventh segment of the male very minute and tuberculate ; intermediate and posterior coxæ approximate ; metasternum convex and slightly transverse ; tarsi thick, first joint very minute, second large, thickened at tip, third smaller, cylindrical, one claw.

In general facies this genus is very different from *Euplectus*, and nevertheless the generic characters are much the same ; the last joint of the maxillary palpi is much larger and a little oblique inwardly. One species only is included in this genus.

Prodalma capensis,
Plate XVI., fig. 19.

Entirely rufous ; antennæ and legs testaceous, covered with a long pilosity not closely set ; head large, convex, and a little broader than long, rounded behind, slightly attenuate in front, with the frontal part truncate, and having two foveæ situated between the eyes as well as two oblique sulci connected with a well-marked, transverse frontal sulcus ; vertex carinate ; antennæ moderately elongate, the two basal joints larger than the others, the second one briefly ovate, third somewhat conical, fourth to eighth inclusive moniliform, seventh and eighth transverse, tenth larger, transverse, eleventh large, briefly ovate, acuminate ; prothorax a little broader than the head and especially longer, very cordate, and having two lateral foveæ and a transverse sulcus slightly narrowed ; elytra not longer than the prothorax and a little broader at apex, slightly rounded laterally and attenuate at base, without shoulders but minutely dentate at the humeral angle, bifoveate at base and with the dorsal stria shortened before the median part ; abdomen slightly more attenuate at base than the elytra, slightly arcuate laterally ; first dorsal segment much larger than the others, impressed transversely at base ; anterior femora thicker than the others.

Male : Antennæ much thicker ; elytra bisinuate at apex ; third to fifth ventral segment short, equal, sixth large, incised at apex, seventh minute, tuberculose ; eyes larger.

Female: Antennæ slender; eyes minute; elytra truncate, straight at apex; third to fifth ventral segment decreasing in size, sixth large, transversely triangular. Length 1 mm.

The seventh ventral ring in the male seems to be reduced to a strong tubercle situated in the centre of a depression at the apex of the preceding segment, but it is, I think, without doubt a true segment.

I have found two examples only: one, a female, in the Platklip Stream at the foot of Table Mountain, and one male at Newlands, also at the foot of Table Mountain.

Gen. TRIMIODYTES.

Subelongate and hardly convex; head large; antennæ distant at base, moniliform, and with an inconspicuous club; palpi moderately elongate, the two basal joints slightly thickened at apex, third minute, fourth large, fusiform, subelongate and acuminate; prothorax cordate, transversely sulcate and trifoveate; elytra little elongate, without shoulders and not dentate, marginal sulcus deficient, sutural stria entire, dorsal one short; abdominal segments equal, the second ventral one larger than the others; intermediate and posterior coxæ approximate; tarsi large, first joint very small, second conical, third cylindrical; a single strong claw.

This genus is very closely allied to *Prodalma*, from which it differs by the more elongate body, the elytra having neither humeral tooth nor subepipleural groove, the first dorsal segment equal in size to the following, and the longer tarsi.

Trimiodytes palustris,
Plate XVI., fig. 18.

Totally chestnut red; the legs, antennæ, and palpi red, with scattered yellowish hairs; elytra darker than the rest of the body; head large, slightly attenuate in the anterior part, frontal part little oblique on each side, vertex slightly convex, briefly and indistinctly carinate, and having two small distant foveæ placed nearly before the eyes, merging into one another and joined in the anterior part by two round sulci; eyes small, median; antennæ elongate, stout, with the first joint bigger than the others, the second ovate and larger than the others, the third suboblong, slightly obconical, fourth to eighth inclusive moniliform, ninth a little bigger and slightly transverse, tenth larger than the preceding, eleventh ovate, truncate at base, acuminate at apex; prothorax a little narrower than the head, strongly cordate, slightly sinuate behind the median part and having deep but not broad lateral foveæ, median one large, shallow, subtriangular, transverse sulcus narrow, and base itself impressed transversely; elytra a little broader and longer than the prothorax, attenuate at base, sides very little rounded, deeply bifoveate at base, dorsal stria not reaching the median part; abdomen almost equal in length to the elytra and attenuate behind; metasternum convex; legs sufficiently long, femora little thickened; tibiæ not quite straight, slightly thickened at apex. Male unknown. Length 1·10 mm.

Hab. Cape Colony (Muizenberg).

GEN. EUPLECTUS, Leach,
Zool. Miscell., iii., 1817, p. 80.

Elongate, sublinear; head large; maxillary palpi moderately long, first joint not conspicuous, second slightly clavate at tip, third minute, fourth ovate, acuminate; club of antennæ triarticulate; prothorax cordate and having three foveæ joined by a transverse sulcus, discoidal fovea unconnected with the others; elytra moderately elongate, bifoveate at base, dorsal stria conspicuous, short; abdomen elongate, first three dorsal segments of same size as the first, fourth much larger, male with seven segments on the under side, female with six; first ventral segment depressed at apex between the coxæ, second to fourth subequal, fifth smaller, sixth much larger than the preceding one, arcuate and hardly ampliate laterally, seventh large, rhomboid, and more or less sulcate or subcarinate longitudinally; posterior and intermediate coxæ approximate; tarsi moderately slender, first joint minute, second elongate, slightly arcuate and hardly thickened, third cylindrical; a single claw.

The distinguishing characters of this genus are: fourth dorsal segments very wide, basal one in the abdominal part flattened in the apical part, sixth larger than the second, arcuately concave and not ampliate laterally, seventh one without operculum in the male.

The genus is largely represented in Europe and North America, and has few representatives in Africa. These insects live, as a rule, in marshy places.

EUPLECTUS DISCOIDALIS.

Elongate, moderately convex, rufous, with the antennæ and legs paler, briefly and sparsely pubescent; head large, attenuate for a short space, frontal sulcus transverse, strong, entire; eyes large, two large foveæ and two straight sulci; vertex briefly sulcate next to the neck; antennæ short, slightly thickened, the two basal joints quadrate, third to eighth inclusive moniliform, ninth slightly, tenth much transverse, eleventh briefly ovate and obtusely acuminate; prothorax of nearly the same width as the head, subcordate, more attenuate behind than in front, obtusely dentate laterally, beyond the median part and alongside a large lateral fovea, median fovea small, transverse sulcus angular, discoidal fovea free, deep, rounded, base bifoveate; elytra elongate with the shoulders quadrate, not attenuate at base, sides slightly rounded towards the median part, basal part with two large foveæ, dorsal stria more or less abbreviated and situated at about one-third of the width; first three dorsal segments of abdomen short, the two basal ones impressed transversely in the middle of the basal part, fourth twice as long; metasternum sulcate; tibiæ slightly thickened externally beyond the median part.

Male: Fourth, fifth, and sixth ventral segments impressed transversely, seventh large, strongly triangular at apex, and having a longitudinal, entire, and slightly arcuate carinule.

Female: Sixth and last ventral segment large, acuminate at tip, last dorsal one very small and acutely dentate at tip. Length 1·40 mm.

This species is closely allied to *E. africanus*, Raffr., from Zanzibar and Abyssinia, and is to be distinguished by sexual characters only; in *E. africanus* male the ventral segments have no impression, the seventh is smaller and more rounded; in the female the last ventral segment is rounded and the last dorsal one not dentate.

Hab. Rhodesia (Salisbury).

EUPLECTUS QUADRICEPS.

Very similar to the preceding species; differs in the shape of the head, which is not at all narrowed in the anterior part, and is thus nearly square; the antennæ are thicker, the tenth joint not so transverse, and the last one is more elongate, nearly straight laterally and is rounded at tip.

The sexual differences are very marked in the male, the fourth ventral segment of which is obtusely angular and projects a little over the following one, which is deeply and transversely impressed under the apical angular edge of the fourth, the sixth is impressed in the centre, and the seventh bears only at tip a small elongate tubercle; in the female the last dorsal segment has no tooth, and the last joint of the antennæ is quadrate.

Hab. Rhodesia (Salisbury).

E. africanus from Zanzibar and Abyssinia, *E. discoidalis* and *E. quadriceps* from South Africa, are so closely allied to one another that they might be taken for one and the same species but for the sexual characters, which differ greatly in the male as well as in two of the females.

Gen. ASYMOPLECTUS.

Elongate, depressed, parallel; head large; antennal club triarticulate; prothorax more or less trapezoid with all the angles rounded, three foveæ connected by a transverse sulcus, the discoidal fovea frequently wanting; elytra subelongate, bifoveate at base and without dorsal stria; abdomen more or less elongate, first two dorsal segments equal, third and fourth larger, principally in the male, which has seven ventral segments, and the female six, the first one between the coxæ is carinate, the second to fourth inclusive subequal, fifth smaller in male, sixth nearly invisible in the middle but conspicuous and irregular laterally, seventh large, carinate; in the female the fifth joint is a little smaller, and the sixth large, subtriangular; the other characters are as in *Euplectus*.

This genus is closely allied to *Euplectus* and *Bibloplectus*; it differs from both by the first ventral segment being strongly carinate from base to apex; in the male the sixth segment is almost entirely hidden in the median part, but wide and more or less irregular laterally; the third and fourth dorsal segments are much larger than the others, while the fourth one is larger in *Euplectus*, and in *Bibloplectus* they are all of about the same size. Owing to the shape of the prothorax and the absence of dorsal stria on the elytra it resembles much more *Bibloplectus* than *Euplectus*.

The genus occurs only in Africa; it includes *Euplectus antennatus*, Raffr., from Abyssinia, and six South African species.

Asymoplectus discicollis, Raffr.,
Plate XVI., fig. 21.

Bibloplectus discicollis. Rev. Entom., vi., 1887, p. 53.

Elongate, piceous black or brown; elytra more or less reddish brown; antennæ and legs testaceous or red, covered with a pale pubescence; head of a moderate size, attenuate in front, and broad between the eyes, bifoveate, and having two not particularly oblique sulci joined with a transverse anterior one, vertex minutely impressed near the neck; antennæ moderately robust, the first two joints larger than the others and the second ovato-quadrate, third to eighth moniliform, ninth to tenth a little larger, increasing in length and transverse, eleventh moderately ovate and obtusely acuminate; prothorax a little larger than the head, suborbicular, conspicuously foveate on each side and having a transverse sulcus angular in the middle and produced behind in a subelongate, small fovea; elytra broader than the prothorax and much longer, parallel laterally, quadrate at the shoulders and subdentate, and having three foveæ at the base, the external one of which is much smaller than the others and now and then more or less sulciform; abdomen longer than the elytra; metasternum convex, sometimes obsoletely impressed; tibiæ slightly thickened externally behind the middle.

Male: Third dorsal abdominal segment nearly twice as large as the preceding one, fourth a little smaller than the third, fourth ventral one obtusely angular in the middle, totally and deeply impressed, fifth much smaller, very angular in the middle, median part of the sixth inconspicuous, but with an arcuate sulcus on the right side, dentate at apex, left side wider, much dentate and incised, seventh with an arcuate longitudinal ridge; intermediate and posterior trochanters slightly angular.

Female: Third and fourth dorsal segments only a little longer than the preceding ones, the fifth ventral one one-quarter shorter, sixth large, triangular, and sometimes slightly impressed. Length 1·10–1·30 mm.

In my original description I mistook the female for the male.

Hab. Cape Colony (Stellenbosch).

ASYMOPLECTUS IRREGULARIS,
Plate XVI., fig. 23.

Rather convex, chestnut brown, with a moderately soft grey pubescence; head small and short; antennæ thicker, club larger and the eleventh joint with a longer point; prothorax longer than broad, more attenuate in the anterior than in the posterior part, and having a nearly straight transverse sulcus much extended behind in the median part; elytra having the sides much rounded and an external fovea briefly sulciform; first dorsal abdominal segment impressed in the middle at base; tibiæ little thickened.

Male: Third dorsal segment nearly three times the size of the preceding one, fourth smaller; fourth ventral one slightly sinuate in the middle and minutely impressed at apex, fifth nearly straight, sixth very small in the median part and with a round incision, right side wide, truncate at apex and strongly bidentate, left side deeply incised at apex, seventh much rounded at apex and having a slightly arcuate longitudinal carina, and near it a slight depression; metasternum hardly impressed.

Female: Third dorsal segment not twice as large as the preceding one, fifth ventral segment shorter by one-half than the fourth, sixth large, subtriangular, rounded at apex and slightly impressed. Length 1·20 mm.

Larger and more convex than *A. discicollis*, and of a much lighter colour.

Hab.   Rhodesia (Salisbury).

ASYMOPLECTUS CAVIVENTRIS,
Plate XVI., fig. 22.

Moderately depressed, black or piceous black, sometimes with the discoidal part of the elytra brown; antennæ and legs red; head smaller than in *A. discicollis*, more elongate than in *A. irregulari*; antennæ as in *A. discicollis*, the last joint is, however, a little longer; prothorax briefly subovate, longer than broad, more attenuate in the anterior than in the posterior part, the transverse sulcus angulate in the middle and extended behind; disk with a sulciform fovea or an abbreviated sulcus more or less obsolete; elytra not much broader than the prothorax, rounded for a short space laterally and having a briefly sulciform, external fovea.

Male: Third dorsal segment of abdomen thrice as broad as the preceding one, fourth a little smaller than the third; fourth ventral segment slightly sinuate at apex and transversely impressed in the middle, fifth incised in the middle in such a way as to form an angle, sixth with the right side sinuate, left side obtusely and widely incised and also obtusely dentate; seventh elongate, carinate, obtusely acuminate at apex; metasternum at times obsoletely impressed; anterior femora much thickened.

Female: Third dorsal segment merely a little larger than the preceding one, the fifth ventral one shorter than the others by one-half, sixth large, triangular, much acuminate at tip.   Length 1·20–1·30 mm.

The distinctive character of this species is the discoidal impressions of the prothorax which are never entirely obliterated.

Hab.   Cape Colony (Cape Town, Newlands, Stellenbosch).

ASYMOPLECTUS LUCTUOSUS,
Plate XVI., fig. 26.

Elongate, somewhat narrow and depressed; piceous black; elytra sometimes with the disk dark brown; antennæ and legs brown, covered with a short, sparse, decumbent pubescence; head large, longer than broad, slightly attenuate in the anterior part; antennæ similar to those of *A. discicollis*, but with a larger club; prothorax not broader than the head, a little longer than broad, nearly straight laterally and with all the angles rounded, nearly evenly attenuate in both the anterior and posterior part; lateral foveæ small, the transverse sulcus not deep, little angular and briefly produced behind.

Elytra a little broader than the prothorax, moderately elongate, parallel laterally with the shoulders quadrate; external fovea large but hardly sulciform; abdomen a little broader behind than at base.

Male: Third dorsal segment not twice as long as the preceding one, fourth nearly equal to the third; of the ventral ones the fourth is hardly sinuate at apex, the fifth deeply foveate in the middle, the sixth is impressed in a circular manner and slightly asymmetrical and obsoletely sinuate laterally, seventh large, triangular, obtusely carinate at apex; metasternum at times inconspicuously impressed.

Female: Third and fourth dorsal segments merely a little larger than the preceding one, fifth ventral one shorter than the fourth by one-half, sixth large, transversely triangular and acuminate at tip. Length 1–1·10 mm.

This species is smaller, more slender and more parallel than the others, and the ultimate ventral segment in the male is less irregular in shape.

Hab. Cape Colony (Cape Town, Stellenbosch).

ASYMOPLECTUS ATERRIMUS.
Plate XVI., fig. 25.

Shorter and broader than the other species, black or piceous black, now and then paler on the disk for a short distance; head large, more attenuate in front than behind, temporal prominences rounded, the foveæ between the eyes larger than the others, the sulci inconspicuous; antennæ shorter and much thicker than in the other species, club less conspicuous; prothorax similar in shape to that of *A. discicollis*, but rather narrower than the head; elytra a little broader than the prothorax, slightly elongate and rounded for a short space; abdomen convex, declivous and much rounded at apex.

Male: Third dorsal segment nearly twice as large as the preceding one, fourth ventral one hardly sinuate at apex and with a minute median tubercle, fifth with the median part broadly depressed transversely and deeply incised circularly, seventh brief, obtuse at tip and with a straight longitudinal carina; metasternum slightly impressed; intermediate and posterior trochanters slightly angular.

Female: Abdomen less convex, third dorsal segment only a little larger than the preceding one, fifth ventral not quite half the size of the preceding sixth transverse, and with the apex obtuse and nearly rounded. Length 1·10–1·30 mm.

This species is shorter and broader than the others, and the ventral segments are symmetrical.

Hab. Cape Colony (Cape Town, Newlands).

ASYMOPLECTUS ATRATUS,
Plate XVI., fig. 24.

Moderately elongate, parallel, depressed, piceous; antennæ and legs testaceous red; head large, attenuate in front; foveæ large and sulci slightly arcuate; antennæ moderately short, joints increasing perceptibly from the third to the apical one and much thickened at apex, while the club, however, is little conspicuous; prothorax slightly cordate and subequal in width to the head; elytra hardly rounded laterally, external fovea sulciform; abdomen moderately short, convex and declivous, obtusely triangular at apex.

Male: Third dorsal segment hardly double the length of the preceding one, fourth much deflexed, the fourth ventral one with a longitudinal carina from end to end, sixth symmetrical, tuberculate in the middle and with the sides simple and oblique, seventh rhomboidal, obsoletely carinate; metasternum convex.

Female, perhaps of this species: Rufous brown (immature); head large; third dorsal segment of abdomen hardly larger than the preceding one, fifth ventral one shorter by more than one-half than the fourth, sixth transverse and rounded at apex. Length 1 mm.

This species is different from the others by its smaller size; the antennæ are shorter, and gradually thickening from the third joint to the tip, so that the club, although of good size, is not so conspicuous. I am not sure, although I believe it, that the female here described belongs to the same species; the head is somewhat larger than in the male.

Hab. Cape Colony (Newlands. I have seen one example only of each sex).

GEN. ANOPLECTUS.

Short, broad, depressed; head large, antennal club hardly conspicuous, triarticulate; prothorax cordate, trifoveate, and having a strong transverse sulcus, longitudinal sulcus not conspicuous; elytra subquadrate, bifoveate at base and without dorsal stria; first to fourth abdominal dorsal segments subequal, first ventral one carinate between the coxæ, second to fourth subequal, fifth minute, sixth small in the middle and ampliate laterally; male with a seventh rhomboidal one, depressed and with a slightly oblique raised line.

This genus has great affinities with *Euplectus*, *Asymoplectus*, and *Bibloplectus*. It differs from *Euplectus* in having the dorsal segments of the abdomen very nearly equal, the basal ventral one strongly carinate, the sixth not wider in the middle than the fifth and no dorsal stria on the elytra; the shape of the dorsal part of the abdominal segments and the much shorter body differentiate it from *Asymoplectus*, and the distinctive characters which separate it from *Bibloplectus*, which has also the dorsal part of the abdominal segment subequal, are the much smaller sixth ventral segment, and in the male the seventh one which has no operculum, a strong characteristic of *Bibloplectus*, the basal joint strongly carinate, the much shorter body, and the decidedly cordate prothorax.

ANOPLECTUS NIGER.

Black, or piceous black; antennæ, palpi, and legs brown (now and again the coxæ, the mouth, and the under part of the abdomen are more or less rufous brown); body covered with a very short, sparse, moderately soft pubescence; head short, attenuate in front with the sides oblique, frontal part with a large, transverse sulcus between the eyes, two strong foveæ and some short sulci; it is slightly sinuate behind near the neck, and impressed in the middle in the shape of an incision; antennæ moderately slender, the two basal joints much larger than the others, the second one ovato-quadrate, third to eighth moniliform, ninth to tenth a little larger than the others and slightly transverse, eleventh briefly ovate and obtusely acuminate; prothorax hardly broader than the head, briefly cordate, lateral foveæ large, median one small, transverse sulcus strong, angular, longitudinal one nearly entire but slender and nearly obsolete; elytra longer than broad and a little broader than the prothorax, subparallel laterally, bifoveate at base, humeral fovea strong, dorsal sulcus wanting, posterior margin fringed with a whitish squamose pubescence; abdomen a little longer than the elytra and somewhat abruptly attenuate at apex; legs moderately robust.

Male: Fifth ventral segment short, sixth subequal in the centre, emarginate triangularly, faintly tuberculate on each side and slightly sinuate, seventh large, depressed, and having a longitudinal, slightly raised oblique line; intermediate tibiæ with a minute apical spur; metasternum more or less depressed.

Female: Metasternum convex, fifth abdominal ventral segment hardly smaller than the preceding one, sixth large, triangular. Length 1–1·20 mm.

Hab. Cape Colony (Stellenbosch).

GEN. XENOGYNA.

Elongate, subparallel, subconvex; head moderately elongate, slightly attenuate in front; eyes moderately large; maxillary palpi moderately long, second joint small, clavate at tip, third small, globose, fourth fusiform, moderately acuminate; antennæ short, thick, intermediate and penultimate joints transverse; prothorax subcordate and with three foveæ, transverse sulci connected with one another, base with several punctures; elytra moderately elongate, subparallel, dorsal stria short; first four abdominal ventral segments subequal, both sexes with six ventral segments, the first one with a triangular depression between the coxæ, the second to the fifth gradually decreasing; prosternum carinate, posterior coxæ near one another.

The facies is that of *Euplectus*, but the prosternum is carinate. Its nearest ally is the North American genus *Eutyphlus*.

XENOGYNA HETEROCERA,
Plate XVI., fig. 17.

Pale ferruginous, and with a brief, pallid pubescence; head longer than broad, slightly attenuate in front, and having two foveæ as well as strong, nearly straight sulci joined with a frontal transverse sulcus, placed behind the eyes, vertex slightly convex, sinuate behind and incised in the middle; antennæ short, thick, first joint subelongate, subcylindrical, second briefly ovate, third globose, slightly transverse, intermediate ones different in each sex, eleventh short, truncate at base, acuminate at apex; prothorax a little broader than the head, subcordate, and having the sides crenulate, lateral foveæ distant from the margin and much larger than the median one, transverse sulcus hardly deep, slightly angulate in the middle, base with several small foveæ; elytra a little broader than the prothorax, much longer than broad, subparallel laterally with the shoulders rounded, bifoveate at base, sutural stria entire, the dorsal one attenuate before the middle; abdomen nearly equal in length to the elytra, first dorsal segment impressed transversely at base, last dorsal one quadrate at apex.

Male: Third joint of antennæ transverse, fourth to fifth twice as broad as the preceding ones, produced inwardly, very transverse, the fifth one thicker than the fourth, sixth to eighth transverse, nearly equal in size to the third one, ninth a little larger, transverse, tenth broader and thicker than the ninth; apical margin of the third ventral segment minutely but acutely bituberculate, sixth transversely depressed at base; metasternum depressed; elytra not attenuate at base and with the shoulders quadrate. Length 1·40 mm.

Female: Fourth to ninth joints of antennæ decreasing in length but increasing in width, tenth a little thicker than the others, club almost inconspicuous; elytra slightly attenuate at base and without any humeral angle; eyes smaller than in male. Length 1·30 mm.

The female is smaller and less robust than the male—an unusual case. Found in mosses.

Hab. Cape Colony (Cape Town).

Gen. RAFFRAYIA, Reitter,
Verh. Naturf. Ver. Brünn, xx. p. 198.

Body subelongate, more or less parallel; head variable: antennæ short and thick, joints pluridentate, intermediate ones sometimes compressed, club hardly distinct; maxillary palpi strong, first joint inconspicuous, second elongate, a little incurved, strongly clavate at apex, third subtriangular, minute, fourth fusiform, large, attenuate at tip and provided with a minute, short, and obtuse appendage: prothorax more or less cordate, foveate and sulcate; elytra with a humeral angle which is most often attenuate, but nevertheless more or less dentate and having a sulcate lateral margin; abdomen broadly marginate, both sexes with six ventral abdominal segments, the first dorsal one larger than the others or equal; posterior coxæ approximate; tarsi triarticulate, the first joint small, the second thicker and subobconic, third subcylindrical and more slender than the others, two claws differing much in size.

This genus was established by Reitter for *Trichonyx antennatus*, Raffr., from Abyssinia. It is a very distinct one, but allied to the European genus *Trichonyx*. Its distinctive feature is the presence of one or two rings of minute tubercles on each antennal joint.

The sexual characters vary in each species, and the eyes, the under part of the head, the epistoma, the elytra, and the inferior segments of the abdomen are subject to modification. These sexual characters vary even in the same species, and there are cases of dimorphism in the male, the more developed ones having the normal characters of their sex, while others less well developed are similar to the female except for the last ventral segment, which displays the usual sexual marks of the male sex; in other cases certain females have the characteristics of the male.

Owing to the antennæ being very often compressed, a very careful examination is needed for ascertaining their true shape.

With the exception of *Raffrayia antennata*, which occurs in Abyssinia, all the other species inhabit South Africa, and are particularly abundant both in number of species and examples in the neighbourhood of Cape Town. They are found under bark and stones, but more especially in sifting the dead leaves of the oak-tree.

Synopsis of Species.

A 2. First dorsal segment of the abdomen much larger than the others.
 B 2. Antennæ slightly clavate, the penultimate joint (more especially the ninth) larger than the intermediate ones ..   *caviceps.*
 B 1. Antennæ without club, the two penultimate joints (especially the ninth) smaller than the intermediate ones.
  C 2. Third joint of the antennæ strongly transverse .. .. ..   *deplanata.*
  C 1. Third joint of the antennæ at least as long as or longer than broad, triangular or globose, never transverse.
   D 2. Prothorax entirely without longitudinal sulcus.
    E 2. Head big and thick, rounded, the sulci shallow, arcuate, the carinate on the vertex obsolete and very short .. .. ..   *incerta.*
    E 1. Head smaller, sides oblique, sulci deep, large and oblique, the carinula of the vertex long and strong.
     F 2. Antennæ more slender, ninth joint globose, tenth scarcely transverse, both much smaller than the intermediate ones   *variabilis.*
     F 1. Antennæ much shorter and thicker, ninth joints slightly and tenth strongly transverse, very little larger than the intermediate ones .. ..   *calcarata.*
   D 1. Prothorax with a longitudinal sulcus more or less complete, but never totally wanting even in the least developed specimens.
    E 2. Prothorax strongly cordiform, as long or nearly as long as broad, longitudinal sulcus more or less obsolete, transverse or angulate in the middle ; shoulders generally attenuated in both sexes.
     F 2. Antennæ more slender, ninth joint globose, colour generally darker, piceous brown .. ..   *armata.*
     F 1. Antennæ much thicker, ninth joint transverse, colour ferruginous, sometimes the shoulders are quadrate in both sexes .. ..   *nasuta.*
    E 1. Prothorax very little cordate, broader than long, longitudinal sulcus complete and well marked, transverse one straight, shoulders very quadrate in both sexes .. ..   *cruciata.*
A 1. First dorsal segment of abdomen not larger than the following ones.
 B 2. Antennæ with the joints (at least the intermediate ones) transverse.
  C 2. Prothorax with a longitudinal sulcus more or less obsolete, and sometimes reduced to an oblong fovea in the anterior part of the base.
   D 2. Prothorax transversely ovate, not cordiform ..   ..   *laticollis.*
   D 1. Prothorax cordiform, at least as long as broad.
    E 2. Longitudinal sulcus deep and well defined .. .. ..   *rugosula.*
    E 1. Longitudinal sulcus more or less interrupted or obsolete.
     F 2. Broad ; antennæ with intermediate joints slightly transverse, ninth and tenth nearly quadrate ; prothorax ampliated on the sides ; elytra slightly longer than wide, ferruginous or testaceous .. ..   *majorina.*
     F 1. More slender ; antennæ with the intermediate joints and also the ninth and tenth very transverse ; prothorax longer, not ampliated on the sides ; elytra longer than broad ; colour generally dark with the feet rufous .. ..   *bicolor.*

C 1. Prothorax without any longitudinal channel, the antebasal
        fovea wanting or round.
D 2. Head without any frontal transverse sulcus.
E 2. Broad and convex; prothorax slightly cordate, broader than
        long; elytra not much longer than wide .. .. *natalensis.*
E 1. Narrow, depressed; prothorax much cordate, longer than
        broad; elytra much longer than broad.
F 1. Larger; head scarcely narrowed in front, sulci deep and
        very oblique; prothorax sinuose on the sides close to the
        transverse sulcus .. .. *pilosella.*
F 2. Smaller; head strongly narrowed in front, sulci fine, little
        arcuated and less distant from each other; prothorax
        regularly cordate without sinuosity on the sides .. .. *abdominalis.*
D 1. Head with a deep transverse sulcus on the frontal part,
        dividing in two the tubercules bearing the antennæ.
F 2. Head large, quadrate; antennæ little compact, ninth joint
        quadrate, tenth very little transverse .. .. *nodosa.*
E 1. Head smaller, longer than wide.
F 2. Ferruginous or rufous; antennæ compact and rather short,
        tenth joint very transverse .. .. *microcephala.*
F 1. Black; antennæ longer and slender, joints third to seventh
        only slightly transverse, eighth to tenth quadrate .. .. *obscura.*
B 1. Antennæ elongate, joints quadrate or even longer than wide *longula.*

RAFFRAYIA CAVICEPS,
Plate XVI., fig. 1.

Elongate, rufous or rufo-ferruginous, with the antennæ and legs paler, covered with a pale pubescence; head slightly attenuate in front, sides oblique, two large foveæ between the eyes and two strong oblique sulci connected in the anterior part, vertex slightly raised and carinate lengthways from apex to base; antennæ rather elongate, club triarticulate, first joint cylindrical, second ovate, third obconic, fourth to tenth transverse, the fifth a little larger than the following ones, while the joints are decreasing from the fifth to the eighth inclusive, ninth and tenth larger than the preceding ones, eleventh briefly ovate, truncate at base, strongly acuminate at apex; prothorax neither longer nor broader than the head, equal in length and breadth, oblongo-cordate, more attenuate in front than behind, lateral foveæ large, longitudinal sulcus obsolete, the transverse one strong, not quite straight; elytra as long as broad, base not attenuate, shoulders well defined, dentate, oblique, sides hardly rounded, dorsal sulcus short; first dorsal segment of abdomen larger than the others, and having two slightly diverging carinulæ reaching to and enclosing one-third of the width of the discoidal part; legs rather elongate and slender; metasternum convex and having a median, minute fovea close to the coxæ.

Male: Under side of the head near the mouth deeply excavated in a quadrate form, posterior part of the upper part having a raised area transverse, trisinuate in the anterior part and carinated longitudinally in the middle; last ventral segment slightly impressed; posterior tibiæ with an extremely minute inner spur. Length 1·60 mm.

Female unknown.

Found under the bark of dead trees.

Hab. Cape Colony (Cape Town, Stellenbosch).

RAFFRAYIA DEPLANATA,
Plate XVI., fig. 5.

Oblong, depressed, rusty red, with the legs testaceous, covered with a rather dense pubescence; head much attenuate in front, somewhat retuse and sinuate close to the neck, median part slightly incised, two minute foveæ placed between the eyes, which are also small, sulci shallow; antennæ short and rather thick, first joint large, subcylindrical, second large, obconic, third to tenth transverse, fifth somewhat larger than the others, ninth and tenth less transverse, eleventh larger than the preceding ones, briefly ovate, truncate at base, nearly cone-shaped, and strongly acuminate; prothorax strongly cordate, broader than the head and with the sides slightly sinuate, lateral foveæ very large, longitudinal sulcus strong, abbreviated in the anterior part, the transverse one also strong and sinuate; elytra little elongate, hardly attenuate at base, rounded but still quadrate, bluntly and minutely dentate, sides slightly rounded, dorsal sulcus broad, deep but short; first dorsal abdominal segment larger than the others and transversely impressed at base; legs moderately short; metasternum convex, simple.

Male unknown.

Hab. Cape Colony (Cape Town). I have seen one example only, found by sifting dead oak-leaves.

RAFFRAYIA INCERTA,
Plate XVI., fig. 4.

Oblong, moderately convex, ferruginous or rufo-testaceous, and with a rather long and dense pubescence; antennæ and legs paler at tip; head short, rounded laterally, rather depressed, attenuate in the anterior part and having behind the eyes, which are very small, two shallow foveæ as well as two light, arcuate sulci, vertex briefly carinate near the neck; antennæ little elongate, first joint large, cylindrical, second ovate, slightly obconic, third transverse, triangular, fourth to eighth compressed, strongly transverse, fifth to eighth decreasing in size, ninth nearly globose, tenth slightly larger than the preceding ones and transverse, eleventh subglobose, nearly conical and strongly acuminate; prothorax larger than the head, cordate, rounded laterally and not at all sinuate behind, lateral foveæ slightly elongate, transverse sulcus strong, slightly arcuate; elytra broader than the prothorax, short, shoulders rounded not angular, hardly dentate, sides rounded, dorsal sulcus short; abdomen longer than the elytra, slightly attenuate at base; first dorsal segment large, impressed transversely at base but not conspicuously; legs moderately strong; metasternum obsoletely impressed near the coxæ.

Male: Last ventral segment larger than the others, broadly but not deeply impressed. Length 1·30–1·40 mm.

This species will be easily distinguished by the rounded head with the sulci thin, shallow, and arcuate, the antennæ decreasing in size from the median to the apical joints, and the intermediate ones which are compressed.

Found by sifting dead oak-leaves.

Hab. Cape Colony (Cape Town neighbourhood).

Raffrayia variabilis,
Plate XVI., fig. 3.

Closely allied to the preceding species and very much like; head convex with the sides oblique, rectilinear, sulci strong, broad, deep, rectilinear, oblique, vertex moderately raised between the sulci and carinate lengthways; antennæ slender, intermediate joints less transverse and compressed than in the preceding species; prothorax narrower, less rounded laterally; elytra a little narrower and less rounded laterally.

Female: Eyes very minute; elytra attenuate at base, without humeral angles and rounded laterally; metasternum convex and simple. Length 1·20 mm.

Male, typical form: Eyes large; under part of head deeply excavate transversely, bottom part of excavation finely and transversely carinate in the anterior part with the posterior edge produced in the middle and carinate; elytra rather elongate, hardly attenuate at base, humeral angle subquadrate and well defined; last ventral segment impressed. Length 1·20–1·30 mm.

Male (var. β): Similar to the type form; head also excavate underneath, but eyes as small as in the female; elytra less quadrate at the humeral angles.

Male (var. γ): Entirely similar to the female; head not excavate underneath, last ventral abdominal segment impressed. Length 1·20 mm.

R. variabilis is very closely allied to *R. incerta*, and it is difficult to ascertain which is the female; the head, however, is distinctly smaller in that sex, the sides are oblique, not rounded, the sulci are much larger and more distinctly marked and also not arcuate; the antennæ are not so broad, and the prothorax neither so broad nor so rounded.

This species exhibits a very peculiar and extremely rare case of polymorphism in the male sex, of which we know three forms, two of which are very distinct from that of the female owing to the large and deep excavation on the under side of the head, but the third form is almost entirely similar to the female, except that the shape of the ventral segments of the abdomen assume the concave form peculiar to the male, and that the last segment is impressed. Such males are hardly distinguishable from the females, unless the penis protrudes.

Very abundant in the neighbourhood of Cape Town (Newlands), together with *R. incerta*. Typical males are not so numerous as the females, and of the two varieties of that sex β and γ appear to be very scarce.

RAFFRAYIA NASUTA,
Plate XVI., fig. 10.

Oblong, somewhat convex, ferruginous, rufous or testaceous, covered with a pale pubescence; legs and last joint of antennæ lighter in colour; head rather large, slightly rounded on the sides, attenuate in front, sulci strong, oblique, vertex with a long carina; eyes minute; antennæ short, thick, first joint shorter than usual, second quadrate, rounded, third slightly transverse, eleventh briefly ovate, subconical, and abruptly acuminate; prothorax cordate, a little broader than the head, convex, slightly sinuate laterally behind, lateral foveæ elongate, median sulcus variable in length, sometimes obsolete, the transverse one strong, angular in the middle; elytra longer than broad, rugosely striate transversely but in a desultory manner, attenuate at base, shoulders absent, dorsal sulcus strong, disappearing in the median part; first dorsal ventral segment large, hardly impressed transversely at base; metasternum convex and simple. Length 1·40 mm.

Female: Head truncate in the anterior part, sulci connected in front.

Male: Frontal part of head slightly produced between the antennal tubercles, anterior sulci not connected; under side of head with a large fovea, much transverse, geminate, and the posterior margin of which is sinuate; last abdominal ventral segment obsoletely impressed.

Var. β. Similar to the type; elytra having in both sexes the shoulders noticeable, a little rounded, but still quadrate; eyes large.

The shape of the elytra vary extremely; in the type form both sexes are attenuated at base without prominent shoulders, and the eyes are small; in the variety β the elytra are not attenuated at the base in both sexes, and the shoulders are very marked, and square, and the eyes are much larger.

Whilst in *R. variabilis* polymorphism occurs only in the male, in *R. nasuta* it is conspicuous in both sexes, but the characteristics of the sexes do not vary, and form β must be considered as a mere variety.

It is easy to distinguish the male of this species by the projecting frontal part and the excavation on the under side of the head.

R. nasuta is closely allied to *R. armata*, both having a longitudinal sulcus in the prothorax, but it differs in being smaller and of a lighter colour, the head is longer, the antennæ stouter, and the prothorax less transverse.

Common in the neighbourhood of Cape Town (Newlands, Table Mountain). The variety β is much rarer, and seems so far to be only found at Newlands.

RAFFRAYIA CALCARATA,
Plate XVI., fig. 6.

Very closely allied to *R. variabilis* and *R. incerta*, and differs merely by the sulci on the head being less oblique, the antennæ thicker and shorter and the joints more moniliform, the tenth is slightly and the ninth very transverse; shoulders oblique, better defined; the impression on the first dorsal abdominal segment is much more conspicuous, and narrower than the third part of the width of the disk.

Male: Metasternum depressed, last ventral segment much sinuate at apex; intermediate tibiæ with a long, strong, oblique inner spur a little before the apex.

Female: Metasternum convex, tibiæ without spur.

Only a few examples found in the vicinity of Cape Town.

The three species *R. incerta*, *R. variabilis*, and *R. calcarata* are very closely allied; the male of *variabilis* is at once distinguished by having the under part of the head impressed; in *calcarata* and in *incerta* the under part of the head is not impressed, but in the former the intermediate tibiæ have a very long spur, which is entirely wanting in the latter.

The females are not so easily distinguished. *R. incerta* differs from both *R. variabilis* and *R. calcarata* in being larger, more depressed and more pubescent; the head is larger and flatter and the sulci more slender and arcuate; the intermediate joints of the antennæ are also more transverse and depressed than in the other two species. *R. variabilis* and *R. calcarata* are very similar; both have the sides of the head linear and oblique, the sulci deep, broad, oblique, not arcuate; the difference between the two is found in the antennæ. In *R. variabilis* they are rather elongate and slender, not compact, each joint having the appearance of being pedunculate, the ninth one is globose, and the tenth slightly transverse, whilst in *R. calcarata* they are shorter, stouter, and more compact, being rather moniliform, and the ninth and tenth joints more transverse; in *R. variabilis* the transverse impression at the base of the first dorsal segment is very feeble, but extends in width to more than half the segment; in *R. calcarata* the same impression is narrower than a third of the width, but it is better defined, as it consists of two foveæ united by a transverse depression.

R. calcarata seems to be rare, whilst *R. incerta* and *R. variabilis* are very abundant.

RAFFRAYIA ARMATA,
Plate XVI., fig. 2.

Oblong, rather convex, ferruginous, brown or piceous; covered with a greyish pubescence; antennæ and legs rufous; head broader than long, attenuate in the anterior part, sides oblique, between the eyes, which are not large, two foveæ, sulci deep, oblique, convex in front, antennal tubercles sulcate traversely, vertex with a long carina; antennæ little elongate, second joint briefly ovate, third very briefly triangular, fourth to eighth transverse, briefly decreasing, ninth subglobose, tenth larger, transverse, eleventh subglobose, nearly cone-shaped and acuminate at tip; prothorax transverso-cordate, broader than the head, longitudinal sulcus shallow, lateral foveæ slightly elongate, transverse sulcus deep with the median part angulate; elytra moderately elongate, slightly attenuate at base, humeral angles oblique, obtusely dentate, dorsal sulcus disappearing before or at the median part; first dorsal segment of the abdomen larger than the others, and more or less impressed transversely at base. Legs robust.

Male: Head with two transverse excavations underneath, the anterior one larger and deeper than the posterior one, which is divided by a longitudinal carina; metasternum and also the last ventral segment broadly impressed; intermediate trochanters slightly mucronate inwardly at base.

Female: Metasternum convex, head simple underneath. Length 1·30–1·50 mm.

This species is found in company with *R. incerta* and *R. variabilis*, but both sexes are easily distinguished from them by the longitudinal sulcus of the prothorax, which is more or less clearly defined but never entirely wanting, and also by the larger size and darker colour. The males are at once differentiated by the impressions on the under side of the head.

Hab. Cape Colony (Newlands—neighbourhood of Cape Town).

RAFFRAYIA CRUCIATA,
Plate XVI., fig. 7.

Oblong, rather convex, ferruginous, clothed with a moderately long flavous pubescence, last joint of antennæ of a paler hue; head attenuate in front, sides oblique, between the eyes, which are moderately large, are two foveæ, the two sulci are little elongate, but oblique, the vertex is carinate; antennæ short, thick, second joint quadrate, third briefly triangular, fourth to eighth strongly transverse and decreasing, ninth slightly transverse, smaller, tenth slightly transverse, larger than the preceding one, eleventh subquadrate at base, cone-shaped at apex and acuminate; prothorax broader than the head, narrower than long and slightly cordate, lateral foveæ sulciform, longitudinal sulcus deep, entire, the transverse one straight; elytra subquadrate, hardly attenuate at base, shoulders nearly quadrate and dentate, vaguely and sparsely rugose; abdomen longer than the elytra, first dorsal segment large, deeply impressed transversely at base, the impression covering a third of the disk.

Male: Metasternum and last ventral segments impressed, the first slightly, the second broadly.

Like *R. armata* and *R. nasuta*, *R. cruciata* has a longitudinal sulcus on the prothorax, but it is deeper, more defined, and similar in size to the transverse one, which is straight, both sulci cutting thus into one another at right angles. The size is larger, the antennæ stouter, the prothorax less cordiform, the elytra shorter, with the shoulders well marked and nearly square.

This species seems rare. I have seen only one pair, which I found under stones near Cape Town.

RAFFRAYIA MAJORINA (female), Raffr.,
Plate XVI., fig. 8.

Rev. Entom., 1887, p. 44; pl. 11, figs. 4, 5.

R. pallidula (male), Raffr., *loc. cit.*, p. 44.

Oblong, robust, ferruginous, rubro-ferruginous or rufo-testaceous, covered with a flavous pubescence, legs rufous; head little attenuate, moderately elongate, two large foveæ behind the eyes and also two slightly arcuate sulci connected in front, antennal tubercles foveate, vertex sinuate and slightly impressed near the neck; antennæ thick, second joint subquadrate, third briefly obconical, fourth to tenth hardly decreasing in width, fourth to eighth large, and ninth to tenth less transverse and subquadrate, eleventh not broader than the preceding one, but much longer, truncate at base, acuminate at apex; prothorax hardly longer than the head but broader, ampliated laterally in a rounded shape at about the median part, constricted and sinuate from there to the posterior part, lateral foveæ strong, as is also the transverse sulcus, which is angulate in the median part, median fovea strong, longitudinal sulcus more or less defaced and abbreviate, sometimes obliterated, intermediate fovea merely sulciform; elytra more or less obsoletely and dispersedly rugoso-punctate, sometimes nearly smooth, longer than broad, shoulders defined, oblique, and dentate, sides slightly rounded, dorsal sulcus disappearing towards the median part; abdomen shorter than the elytra, first dorsal segment not longer than the others, the two carinules very diverging and little distant at base.

Male: Posterior trochanters with an inward, short, compressed, incurved tooth, intermediate and anterior ones slightly angulate at base, intermediate tibiæ with a short apical spur; last ventral abdominal segment strongly impressed, and slightly tuberculate in the middle of the base; metasternum impressed.

Female: Last ventral segment of abdomen compressed on each side and having a small, horn-like process at apex, last dorsal one obtusely tuberculate at tip; metasternum convex. Length 1·90–2 mm.

This species varies much in colour, the longitudinal sulcus of the prothorax is also variable, being at times entirely absent, or reduced to an oblong, median fovea. Not having had at first a long series of examples to examine, I was led through these variations to believe that there were two different species, *i.e.*, *majorina* and *pallidula*; but I have been able since to examine more specimens, and I have ascertained that the two are only one species.

With this species begins the group in which the first dorsal segment of the abdomen is not larger than the following ones.

Hab. Cape Colony (environs of Cape Town—Rondebosch). Rare. Found under bark at foot of dead pine-trees.

RAFFRAYIA BICOLOR,
Plate XVI., fig. 14.

Rather elongate and parallel, little convex, piceous or rufous brown; elytra sometimes red, and antennæ and legs rufous, covered with a sparse, decumbent pubescence; head rather elongate, hardly attenuate in front, somewhat sinuate laterally, two foveæ behind the eyes, sulci slightly arcuate and joined in front, vertex briefly carinate; antennæ strong, first joint elongate, cylindrical, second subquadrate, third very briefly subtriangular, fourth to eighth much transverse, compressed, ninth to tenth less transverse, eleventh moderately short, subrotund at base, acuminate at tip; prothorax not longer than the head, but a little broader, cordate, narrowed and sinuate past the middle, lateral foveæ large, transverse sulcus little angular, longitudinal sulcus more or less attenuate in front and behind; elytra rather elongate with the sides subparallel, the shoulders oblique and dentate, the base with three foveæ, dorsal sulcus ending abruptly in the median part; abdomen rather elongate, a little narrowed at base, first dorsal segment not larger than the others, narrowly but deeply impressed at base.

Male: Third ventral segment impressed on each side and fasciculate, last one deeply impressed. Length 1·50 mm.

This species is easily distinguished by its elongate and parallel form, the strong antennæ, and the longitudinal sulcus of the prothorax which may be reduced to a longitudinal fovea on the disk. The colour is very variable.

Rare. I have found two specimens only, in mosses.

Hab. Cape Colony (Cape Town).

RAFFRAYIA RUGOSULA, Raffr.,
Rev. Entom., 1887, p. 46.

Subelongate, rufous brown or ferruginous, covered with a yellowish pubescence; head and prothorax densely subrugose; elytra more or less vaguely punctate at base; head slightly attenuate in front, bifoveate between the eyes, with the two sulci little oblique and strong, vertex carinate; antennæ short, thick, first joint large, subcylindrical, second very little smaller, subquadrato-ovate, third to ninth moniliform, slightly transverse, ninth a little smaller, tenth transverse, a little larger, eleventh larger, briefly ovate and with the apex somewhat cone-shaped; prothorax cordate, a little longer than the head, much broader, rounded laterally before the median part, sinuate past the middle, trifoveate, lateral foveæ large; transverse sulcus in the shape of a comma, longitudinal sulcus median, strong, attenuate in front, base with two foveolæ; elytra a little longer than broad, subparallel laterally and not attenuate at base, shoulders oblique, dentate, dorsal sulcus reaching the median part; first dorsal segment of abdomen a little shorter than the preceding ones, and with a moderately deep transverse impression at base; head rugoso-punctate underneath.

Male: Anterior and intermediate trochanters briefly dentate, intermediate tibiæ with a short apical spur; last ventral segment of abdomen totally and deeply concave and trisinuate at tip. Length 1·50–1·80 mm.

Not so broad as *R. laticollis*, the prothorax is not transverse, and the punctuation on the head and prothorax is very striking.

Very rare.

Hab. Cape Colony (Stellenbosch).

RAFFRAYIA LATICOLLIS,
Plate XVI., fig. 16.

Oblong, rather broad, rufous or rufo-ferruginous, covered with a longer and denser yellowish pubescence; head rather large, attenuate in front, sides oblique, behind the eyes, which are rather large, are two large foveæ and two oblique sulci, vertex carinate; antennæ short, thick, second joint ovate, third very briefly triangular, fourth to eighth transverse, of nearly the same size, ninth much less transverse, tenth transverse, eleventh subquadrate at base, slightly cone-shaped at tip and acuminate; prothorax much broader than the head, and also much broader than long, rounded laterally, narrowed behind the median part and slightly sinuate, lateral foveæ broad and slightly sulciform, transverse sulcus angulate in the middle, median fovea minute, and longitudinal sulcus shortened in the anterior part; elytra longer than broad, not attenuate at base, shoulders oblique and dentate, sides slightly rounded, dorsal sulcus disappearing in the median part; abdomen shorter than the elytra, segments short, the first one narrowly impressed at base. Length 1·60 mm.

Male: Antennæ a little longer and more slender; intermediate tibiæ with an inner, moderately obtuse spur, posterior ones ciliate and with a sharp apical spur; last ventral segment totally and strongly depressed transversely, last dorsal segment large and convex; metasternum impressed.

Female: Last ventral segment impressed in the middle and on each side.

Easily distinguished by its massive shape, and the broad, nearly transverse prothorax, the longitudinal sulcus of which is well marked, but ends abruptly a little after the median part.

Apparently rare.

Hab. Cape Colony (neighbourhood of Cape Town, Stellenbosch).

Raffrayia abdominalis.

Elongate, somewhat depressed, rufo-testaceous, with the antennæ and legs testaceous, briefly but densely pubescent; head attenuate in the anterior part, somewhat rounded laterally, and having two sulci slightly arcuate and near to one another, ending behind in minute foveolæ, and in front in a single, large frontal fovea; eyes small, placed on the under part of the head; antennæ rather elongate, first joint elongate, cylindrical, second subquadrate and longer than broad, third to seventh strongly transverse and slightly increasing, especially lengthways, eighth smaller and transverse, ninth slightly transverse, tenth more transverse, eleventh cone-shaped; prothorax larger than the head, cordate, and having on each side a large fovea and a minute, transverse sulcus very narrow in the median part; elytra much longer than broad, subparallel laterally; dorsal sulcus obsolete and attenuate towards the middle; abdomen a little shorter than the elytra, dorsal segments subequal.

Male: The second basal segment of the dorsal part of the abdomen is a little larger than the others, densely and briefly covered with a subsquamose pubescence, apical margin a little angulate and fasciculate along the median part, third segment shorter by one-half, deeply incised in the middle, a little excavate, the bottom of excavation filled with fascicles of long hairs; metasternum slightly depressed, last ventral segment very much depressed. Length 1·10 mm.

Closely allied to *R. pilosella*, but the head is smaller and narrower in front, the size is also smaller; the shape of the abdomen in the male is very peculiar.

Hab. Cape Colony (neighbourhood of Cape Town).

Raffrayia pilosella,
Plate XVI., fig. 15.

Elongate, subparallel and depressed, rufo-ferruginous, densely but briefly pubescent; antennæ and legs rufous; head large, subquadrate, a little attenuate in front, and having two oblique sulci, vertex finely carinate; eyes large; antennæ slender and long, first joint elongate and cylindrical, second subquadrate, third briefly triangular, fourth to eighth slightly transverse and a little decreasing in length, ninth subglobose, hardly transverse, tenth a little broader and transverse, eleventh short, subconical and acuminate; prothorax a little broader than the head, strongly cordate, sinuate laterally past the middle, lateral foveæ sulciform, median one minute, transverse sulcus strong, angulate in the median part; elytra elongate, hardly attenuate at base, shoulders moderately rounded and obtusely dentate, dorsal sulcus evanescent towards the median part; first dorsal segment of abdomen transversely and moderately deeply impressed at base, the impression narrower than a third of the disk; legs robust.

Male: Metasternum flattened, last ventral segment transverse, strongly impressed, posterior tibiæ with a brief but sharp spur at apex. Length 1·50–1·60 mm.

Easily distinguished by the large, nearly square head, the antennæ more slender than in the other species, the very cordate prothorax nearly dentate laterally close to the lateral fovea, the long elytra, and the much thicker, but short pubescence.

RAFFRAYIA NATALENSIS,
Plate XVI., fig. 9.

Oblong, rather broad, chestnut red, covered with a pale pubescence; head small, much attenuate in front, and having between the eyes, which are large, two foveæ and two slightly arcuate sulci joining before the depressed forehead; antennæ moderate, a little clavate at apex, second joint quadrate, third ovate, fourth to eighth subquadrate and transverse, ninth quadrate, tenth a little larger, subquadrate and transverse, eleventh large, truncate at base, slightly cone-shaped and acuminate; prothorax much larger than the head, nearly as broad as long, cordate, somewhat deeply sinuate past the median part near the lateral foveæ, which are wide and deep, transverse sulcus deep, angulate in the middle; elytra a little longer than broad, slightly rounded laterally, hardly attenuate at base, shoulders oblique, obtusely dentate, dorsal sulcus evanescent towards the median part; abdomen almost longer than the elytra, first segment with two large carinules diverging much, and inclosing the fourth part of the width of the disk. Female. Length 2·10 mm.

One of the largest species of the genus. In general appearance it resembles *R. rugosula*, but differs much on account of the head being smaller and the antennæ somewhat clavate at tip; the prothorax is not so wide, and the longitudinal sulcus is totally wanting.

Hab. Natal (Escourt). One example only.

RAFFRAYIA NODOSA,
Plate XVI., fig. 11.

Oblong, rather thick, ferruginous, covered with a flavous pubescence, legs and apical part of antennæ rufous; head small, quadrate, nodose on each side in front, and having between the eyes, which are large, two foveæ, sulci hardly joined in the anterior part, transverse frontal tubercle deep, antennal tubercles oblique laterally, vertex subconvex, simple, neck with a hardly perceptible carina; antennæ somewhat elongate, first joint cylindrical, second subquadrate, third to eighth slightly transverse and little decreasing in size, ninth to tenth subquadrate, last one moderately elongate, somewhat cone-shaped and acuminate at tip; prothorax much larger than the head, more attenuate in front than behind, rounded laterally in the middle, lateral foveæ slightly elongate, median one deep, minute, transverse sulcus deep, angulate in the middle; elytra hardly as long as broad, not attenuate at base, shoulders oblique, well defined and dentate, sides slightly rounded, dorsal sulcus evanescent in the middle, first dorsal segment not larger than the others, impressed between two short carinules, little diverging, and inclosing a third of the discoidal width.

Male: Metasternum impressed longitudinally, and the last ventral segment impressed transversely at tip; posterior trochanters obtusely angular, intermediate ones briefly spinose near the base. Length 1·50–1·60 mm.

Head small, square, with the antennal tubercles robust and divided in two by a deep transverse groove; the anterior part of the tubercle obliquely cut laterally, the frontal part is abruptly narrowed.

Hab. Cape Colony (neighbourhood of Cape Town—Newlands).

Raffrayia microcephala,
Plate XVI., fig. 12.

Oblong, rather thick and convex, ferruginous or rufo-ferruginous, covered with a somewhat dense yellowish pubescence; legs and antennæ red at apex; head and prothorax sometimes closely rugoso-punctate, sometimes smooth; head small, longer than broad, slightly attenuate in front, and having between the eyes, which are large and placed behind the middle, two foveæ, the sulci are a little oblique, hardly joined in front, the antennal tubercles are robust and divided by a transverse sulcus, the vertex is carinate; antennæ robust, first joint elongate, cylindrical, second quadrate, third to tenth transverse, subequal, except the eighth, which is somewhat smaller, last one short, somewhat cone-shaped and abruptly acuminate; prothorax much larger than the head, cordate, subconvex on the disk, lateral foveæ strong, elongate, median one minute, deep, transverse, sulcus deep, angulate in the middle; elytra hardly longer than broad, not attenuate at base, shoulders oblique, noticeable and dentate, sides little rounded, dorsal sulcus strong, ending in the median part; first dorsal segment of abdomen strongly impressed between two more or less short carinæ, not much divergent and inclosing one-third of the disk.

Male: Metasternum and also last ventral segment broadly impressed.

Female: Metasternum convex, last dorsal segment minutely tuberculate at apex. Length 1·40–1·60 mm.

Closely allied to the preceding species, but the head is longer, slightly narrowed in front, with the tubercles much smaller; the prothorax is cordiform and very variable; in some specimens the head and prothorax are coarsely punctured, in others they are smooth.

Not rare in the neighbourhood of Cape Town and Newlands. The examples from Newlands have generally the head and prothorax punctured, and those from Cape Town smooth.

Raffrayia obscura.

Somewhat elongate, not much thickened, black, with a rather dense but brief pubescence, a few of the hairs dispersed, long and recurved; legs and antennæ piceous, last joint of the former lighter, tarsi and palpi testaceous; head small, elongate, slightly attenuate in front, two foveæ between the eyes, which are large, sulci oblique, antennal tubercles divided by a strong, transverse sulcus, vertex somewhat convex, briefly carinate close to the neck; antennæ rather elongate, a little thickened but not at all at apex, first joint elongate, cylindrical, second quadrate, third to seventh slightly transverse, eighth to tenth subquadrate, eleventh middling, subturbinate and acuminate; prothorax much broader, hardly longer, cordate, lateral foveæ large, median one minute, transverse sulcus strong, angulate, impressed at base; elytra large, shoulders well defined, oblique, obtusely dentate, slightly rounded laterally, strongly bifoveate at base, dorsal sulcus shortened in the middle; abdomen a little smaller than the elytra, dorsal segments equal and having the carinæ short, inclosing only the fourth of the length of the disk, and having between them a deeply impressed space.

Male: Head subgibbose underneath and slightly impressed on each side before the eyes; metasternum deeply sulcate longitudinally, last ventral segment slightly impressed. Length 1·50 mm.

Resembles much *R. microcephala*, but differs by the darker colour, the longer head, the more slender antennæ and the shorter and more approximate abdominal dorsal carinulæ.

Hab. Cape Colony (neighbourhood of Cape Town—Muizenberg Vlei). One male only.

RAFFRAYIA LONGULA, Raffr.,
Plate XVI., fig. 13.
Rev. Entom., 1887, p. 45.

Elongate, parallel, ferruginous red, shining, briefly and sparingly flavo-pilose, antennæ and legs paler; head moderately elongate, slightly attenuate in front, and having two large foveæ placed past the eyes, sulci little oblique, shallow, vertex carinate, antennal tubercles incised laterally; antennæ slender and elongate, first joint robust, second quadrate, third suboblong, fourth to tenth quadrate, hardly increasing in width, ninth a little longer, eleventh moderately elongate, truncate at base; prothorax strongly cordate, broader than the head, rounded laterally, emarginate past the middle and narrowed, from there slightly sinuate; lateral foveæ large, deep, median one smaller, transverse sulcus angulate; elytra elongate, not much broader than the prothorax, parallel laterally, shoulders oblique, well defined and minutely dentate, dorsal sulcus a little longer than the others, the two carinæ very short and inclosing hardly the third part of the length of the disk.

Male: Head broadly and deeply impressed laterally on each side underneath; ventral abdominal segments deeply impressed transversely, the sixth impressed in an oblong shape and sinuate at apex, third to fourth with a sharp point on each side, trochanters with a very short tooth at base, posterior ones armed with a slightly sinuate, apical spine, intermediate tibiæ briefly spurred at apex.

Female unknown. Length 2·40 mm.

This species is the largest of the genus, and is easily distinguished by its long and parallel form and slender antennæ.

Hab. Cape Colony (Stellenbosch). Seen one example only.

GEN. DALMINA, Raffr.,
Rev. Entom., vi., 1887, p. 46.

One constant characteristic, but an important one, distinguishes this genus from *Raffrayia*, viz., the joints of antennæ have no spines or tubercles, the body is generally longer and more slender, the intermediate joints of the antennæ are in both sexes always a little larger than the others, and they have sometimes a distinct median node.

So far as is known the genus *Dalmina* is exclusively a South African one.

DALMINA GLOBULICORNIS, Raffr.,
Rev. Entom., 1887, p. 47, pl. ii., figs. 6 & 7; Rev. Entom., 1890, pl. iii., fig. 40.

Oblong, piceous black, covered with a somewhat flexible greyish pubescence, elytra red, legs and antennæ rufous, and in immature specimens entirely testaceous; head moderately short, attenuate in front, sides and frontal part punctulate, two foveæ between the eyes which are large, sulci little oblique and connected in the frontal part which is depressed, vertex subconvex, simple; first joint of antennæ rather large, second subquadrate, third ovate, fourth to fifth variable in both sexes, sixth to eighth subquadrate and transverse, ninth to tenth equally long but a little narrower and hardly transverse, eleventh not thicker than the preceding one, merely twice as long and acuminate at tip; prothorax longer than the head and a little broader, cordate, and having two lateral foveæ, transverse sulcus distinctly angular; elytra moderately elongate, slightly rounded laterally, dorsal sulcus strong, attenuate before the median part; abdomen shorter than the elytra, moderately convex, and declivous at apex, first dorsal segments subequal; metasternum foveate in the centre close to the posterior coxæ; legs moderately long.

Male: Fourth and fifth joint of antennæ much larger than the others and forming a large node, the fifth joint being, however, the larger of the two and bi-impressed inwardly; elytra elongate, shoulders oblique and developed, trochanters of the forelegs with a basal spine moderately long and slightly recurved; anterior and intermediate tibiæ with a short spur, the spur longer in the posterior ones; metasternum broadly impressed, last ventral segment deeply impressed.

Female: Fourth and fifth joint of antennæ a little more robust than the following ones, fifth a little larger than the fourth; elytra shorter, attenuate at base, shoulders almost wanting; metasternum convex. Length 1·50–1·60 mm.

The colour of this species in both sexes, and also the shape of the fourth and fifth joints of the antennæ in the male, makes this species very conspicuous.

Hab. Cape Colony (Cape Town, Newlands, Stellenbosch).

DALMINA GRATITUDINIS.

Oblong, subparallel, castaneous, more or less infuscate or pallid, legs, antennæ, and palpi rufous, body covered with a sparse, flexible greyish pubescence; head attenuate in front, temporal angles rounded, two foveæ between the eyes and two anterior sulci joined in the frontal part; antennal tubercles conspicuous, antennæ strong, differing in each sex, last joint testaceous; prothorax cordate, larger than the head and sinuate past the middle owing to the presence of a lateral fovea; transverse sulcus very angular; elytra attenuate at base, shoulders oblique, sides slightly rounded, base bifoveate, sutural stria entire, discoidal sulcus broad but short, the three first abdominal segments decreasing very little in length.

Male: More parallel than the female, and paler, disk of elytra redder; first joint of antennæ a little larger than the others, conical, second ovate, third nearly transverse, fourth a little longer than the preceding one, slightly produced inwardly, sharp underneath, fifth largest of all, transverse, irregularly rounded inwardly, produced outwardly and obtuse, sixth to tenth quadrate, slightly transverse, each one of them slightly larger than the third, last one hardly broader than the penultimate one, but longer, ovate, truncate at base and acuminate at tip; elytra nearly twice the length of the prothorax, less attenuate at base than in the female, shoulders oblique and very well defined; metasternum plane, foveate at apex, last ventral segment much impressed in an oblong shape; posterior tibiæ with an apical spur.

Female: More attenuate in front, darker chestnut red all over; first three joints of antennæ similar to those of the male, fourth trapezoidal, slightly transverse, not longer than the preceding one, but broader, fifth of the same shape as the preceding one but nevertheless a little larger, sixth to tenth smaller, slightly transverse, ultimate one as in the male; elytra shorter, attenuate at base, and almost without shoulders; metasternum slightly convex, minutely foveate; legs simple. Length 1·60–1·80 mm.

Allied to *D. globulicornis*, which has also a node in the median part of the antennæ, but whereas the node in this species is formed by the dilatation of the fourth and fifth joints, in *D. gratitudinis* it is the fifth joint alone which is so dilated.

Hab. Cape Colony (Stellenbosch—banks of the river. February).

DALMINA CONCOLOR.

Oblong, thick, entirely chestnut brown, covered with a brief, greyish pubescence, antennæ and legs rufous; head short and attenuate in front, broad, slightly transverse, two foveæ not much apart from one another, sulci straight and coalescing in the frontal part, which is a little depressed, posterior part a little retuse; antennæ of moderate size, first joint large, second quadrate, third obconical and a little narrower, fourth quadrate, fifth quadrato-elongate and a little larger, sixth and seventh similar to the fourth, eighth quadrate and a little smaller, ninth to tenth quadrate, a little larger than the seventh, eleventh a little thicker than the preceding one, ovate, truncate at base and sharply acuminate at tip; prothorax longer than the head and a little broader, nearly as broad as long, rather short, strongly cordiform and having two lateral foveæ and a much smaller median one, transverse sulcus slightly angular; elytra not much elongate, attenuate at base, no shoulders, dorsal sulcus strong at base but disappearing towards the median part; abdomen longer than the elytra, little convex and attenuate at tip; metasternum simple; legs of moderate size. Female: Length 2 mm.

Differs from the female of *D. globulicornis* by its larger size, the shorter and broader head, the more slender antennæ and the colour.

Hab. Natal (Frere).

DALMINA IRREGULARIS.

Oblong, more slender than the others, piceous, elytra and antennæ brownish, the latter paler at tip, covered with a sparse greyish pubescence, legs rufous brown; head of moderate size, a little longer than broad, attenuate in front, and having between the eyes two foveæ, sulci oblique joining in the depressed frontal part, vertex convex; antennæ somewhat elongate, first joint large, second quadrate, third a little narrower, obconical, fourth quadrato-elongate, fifth also quadrato-elongate but larger, slightly produced inwardly, sixth smaller by one-fourth, and quadrate, seventh and eighth quadrate and larger, ninth and tenth equally broad but shorter and transverse, eleventh not broader, rather elongate, obtusely acuminate and slightly sinuate outwardly; prothorax larger than the head, a little longer than broad, cordate, and having three foveæ, the transverse sulcus is strongly angular, and the very base is impressed; elytra rather elongate; shoulders oblique, distinct, dorsal sulcus deep and disappearing towards the median part; abdomen longer than the elytra and a little narrower, somewhat convex; metasternum much impressed; legs of moderate size, trochanters of the posterior ones with a short spine directed backwards, anterior and intermediate tibiæ with a short spur, last ventral segment of abdomen very much impressed. Male. Length 2 mm.

This species, the female of which is unknown, is greatly differentiated from *D. concolor* by the head and prothorax which are longer than broad as well as by its more slender shape. It resembles more *D. globulicornis*, but the shape of the fourth and fifth joints of the antennæ are not dilated in a node as in the last-named species.

Hab. Natal (Frere).

DALMINA ELEGANS,
Plate XVII., fig. 1.

Elongate, slender, ferruginous, covered with a flexible, sparse, flavous pubescence, tibiæ, tarsi, and palpi, as well as the tip of the antennæ, paler; head hexagonal, equally attenuate in front and behind, moderately depressed, bifoveate behind the median part, the two sulci slightly arcuate, vertex with a long carinule; eyes of moderate size, median and prominent; antennæ elongate, first joint cylindrical, second briefly ovate, third to eighth decreasing, having nearly the same shape but less transverse, and serrate inwardly, third one transversely ovate, and all attenuate at base and apex, ninth and tenth subconical, ninth rather elongate and narrower, tenth shorter and thicker, eleventh hardly larger, suboblong, truncate at base, acuminate at tip; prothorax of nearly the same size as the head, very cordate, with the sides sinuate behind the median part, longitudinal sulcus strong, shortened in front and behind, the transverse one hardly angular, lateral foveæ strong; elytra little elongate, attenuate at base, no shoulders, dorsal sulcus very short; abdomen a little longer than the elytra, rather convex, first dorsal segment slightly larger than the others, and transversely impressed at base; legs elongate, slender; head irregular underneath, impressed in a subtriangular shape in the middle, and having in each side of the anterior part a squamose, depressed area, it is nodose in the median part near the neck, fasciculate and broadly foveate on each side; metasternum convex; last ventral segment impressed and much sinuate at apex. Male. Length 2 mm.

This species differs much from the others owing to the peculiar shape of the head, the much longer antennæ, the joints of which are isolated from one another, the longitudinal sulcus of the prothorax, and the long and slender legs.

Hab. Cape Colony (neighbourhood of Cape Town, Newlands).

Tribe BATRISINI.

The distinctive characters of the tribe are as follows: Median and hind coxæ globular and not prominent, the latter somewhat depressed and triangular, approximate or very little distant from each other; first ventral segment of abdomen concealed under the metasternum or the coxæ, or, when the latter are not quite approximate, looking like a small notch; two very unequal claws to the tarsi.

The other parts of the body are most variable, but it can be said that as a rule in the African insects the body is rather elongate and somewhat cylindrical, the antennæ very distant at base, and the abdomen more or less immarginate, the first segment only having a carinule, which in some cases is entirely wanting.

Some species from Asia and America have a very globular body the abdomen is nearly entirely marginate, and the frontal part prolonged in a long and narrow tubercle bearing antennæ geniculate like those of *Curculionidæ*.

This tribe is perhaps the most numerous of the family, and is largely represented in the tropical parts of Asia and America; they are rare in Europe, in Australia, and in Africa, and so far only four species are known as occurring in South Africa.

One of these four belongs to a genus exclusively African, and met with in Eastern Africa from Abyssinia to Zanzibar; another is, strange to say, the only representative in the Old World of a genus numerous and widely distributed in America. No explanation of this is possible, but there can be no doubt whatever as to the close relationship of this African insect to its American congeners; the third one belongs to a new genus also more closely allied to an American genus than to any other; the fourth belongs to a genus very numerous in the Indo-Malayan region, but having only very few representatives in Africa.

Gen. BATOXYLA.

Elongate, cylindrical; head equal, depressed, attenuate, retuse behind; eyes large, placed backwards; antennæ distant at base, thick, moniliform, club large, triarticulate; prothorax cordate and without sulci; elytra elongate, without striæ, shoulders oblique and well developed; abdomen shorter than the elytra, immarginate, first dorsal segment much larger than the others; posterior coxæ distant, second ventral segment longer than the following ones, first ventral segment conspicuous; legs hardly elongate and rather thick.

This new genus is very closely allied to *Batoctenus*, Sharp, from Central America, and differs only by the head being attenuate in the frontal part, and not at all nodose above the insertion of the antennæ, which are stouter and have a large club; the elytra are without striæ, whereas *Batoctenus* has three; the first dorsal segment of the abdomen is much larger, and the abdomen is entirely without margins.

Batoxyla punctata,
Plate XVII., fig. 4.

Ferruginous, subopaque, totally covered with large, ocellate, but shallow punctures, briefly and sparsely pubescent, the pubescence pallid; head depressed slightly on the upper part and having two minute foveæ between the eyes on the vertex; first joint of antennæ strong, second subquadrate, third to eighth moniliform, slightly transverse, club large, ninth and tenth trapezoid, subtransverse, eleventh truncate at base and a little narrower than the preceding one, thickened at apex and obtusely acuminate; prothorax larger than the head, cordate, foveate on each side past the median part; elytra much longer than broad, bifoveate at base, shoulders oblique, subdentate, sides subparallel, no striæ; abdomen shorter than the elytra, first dorsal segment much larger than the others, tri-impressed at base and bicarinate in the middle; metasternum depressed, last ventral segment broadly but not deeply foveate. Male. Length 1·90 mm.

Hab. Zambezia (Salisbury). Two examples.

Gen. BATRISUS, Aubé,
Mag. Zool., 1833, p. 45.

This genus, *sensu proprio*, has no representative in Africa.

Sub-Gen. TRABISUS, Raffr.,
Rev. Entom., 1890, p. 110; *loc. cit.*, 1894, p. 230.

Body oblong, head more or less trapezoidal, eyes median; antennæ distant at base and with eleven joints; maxillary palpi moderate, first joint inconspicuous, second slightly elongate, third minute, fourth elongate, fusiform, but at times falciform; prothorax cordate, trisulcate lengthways; first dorsal segment of abdomen large, unicarinate only on the side of the base, the other segments immarginate.

This genus differs from *Batrisus* proper by the much more slender form of the ultimate joint of the maxillary palpi, and the first dorsal abdominal segment having one simple and short carina at base, the abdomen being otherwise immarginate laterally.

This sub-genus, which includes about nine species, is exclusively African.

Trabisus dregei, Aubé.

Batrisus dregei, Aub., Ann. Soc. Ent. France, 1844, p. 82.

Oblong, chestnut brown, clothed with a yellowish pubescence; head broad, transverse, and having in front a large transverse impression joined laterally to two foveæ; vertex raised, transverse, punctate behind, briefly carinate; last joint of palpi elongate, fusiform, sinuate and slightly falcate; antennæ robust, joints subcylindrical oblong, eighth small, ninth and tenth larger, eleventh much longer, acuminate; prothorax cordate, a little narrower than the head, median longitudinal sulcus nearly absent, lateral foveæ joined by a transverse sulcus with a large, cruciform, median fovea, base bifoveate on each side; elytra subquadrate, somewhat convex, shoulders high, slightly oblique and obtusely carinate, rounded laterally, trifoveate at base, sutural stria entire, dorsal one short; first dorsal segment of abdomen nearly twice as long as the following one, and very briefly unicarinate at base; legs strong, femora thickened, tibiae slightly incurved; metasternum impressed and sulcate; head broadly excavate in a subquadrate form, the margin of the excavation sharply tuberculate close to the median part of the mouth; intermediate trochanters thickened at apex, last ventral segment impressed transversely. Male. Length 3·20 mm.

This description is made from Aubé's original type, formerly in Reiche's collection and now in mine. Aubé mentioned (Ann. Soc. Ent. France, 1844, p. 82) another example belonging to Schaum's collection, but no other has been recorded from that time.

This extremely rare insect was discovered at the Cape by Drege, but the exact locality is unknown, and I am not aware that it has been met with again.

Gen. ARTHMIUS, Le Conte,
Bost. Journ., vi., p. 91.
Raffr., Rev. Entom., 1894, p. 231.

Sub-Gen. SYRBATUS, Reitt.,
Verh. Naturf. Ver. Brünn., xx., p. 205.
Raffr., Rev. Entom., 1894, p. 231.

Body suboblong, rather convex; head quadrate or transverse; eyes placed behind the median part; last joint of palpi ovate, acuminate; antennae much separated; prothorax cordate, sulcate longitudinally on the sides, median sulcus always wanting, and the transverse one placed before the base; elytra quadrato-elongate and without shoulders in the female and with more or less prominent ones in the male, dorsal stria wanting; abdomen with a rather abrupt declivity behind, broad, marginless, first dorsal segment very large.

This sub-genus differs only from *Arthmius* by having two longitudinal sulci on the sides of the prothorax, whereas there are no sulci in *Arthmius*; this difference is really not important.

Arthmius, as well as *Syrbatus*, was formerly considered as belonging to the genus *Batrisus*, to which it is indeed very closely allied, but the old genus *Batrisus* has now become so large and consists of so many heterogeneous elements that it has become necessary to divide it into several distinct genera, from among which *Arthmius* is one of the best characterized. It differs from *Batrisus* in the shape of the head, which is more quadrate; the eyes are always placed beyond the median part; the prothorax has no median sulcus; the elytra no discoidal striæ; the abdomen is entirely without margin, and the first segment is very large, and the others abruptly declivous.

The head of the male bears generally some foveæ, spines, or tubercles; the shape of the elytra is different from that of the female, the shoulders being more or less prominent in the male, and rounded and obsolete in the female.

Both *Arthmius* and *Syrbatus* are largely represented in America, and the finding of this genus in South Africa was quite unexpected and is of great interest.

SYRBATUS MASHUNA.
Plate XVII., fig. 3.

Oblong, moderately convex, chestnut brown, antennæ rufous at apex, likewise the legs, body covered with a brief fulvous pubescence; head large, quadrato-transverse, sinuate laterally before the middle, frontal part truncate perpendicularly, nearly completely excavate, the excavation trilobate and briefly setose at bottom, with the posterior lobe much larger than the others, frontal part delicately and sinuously sulcate, and briefly carinate lengthways; antennæ very distant at base, third to seventh joints oblong, slightly decreasing in length, eighth quadrate, ninth to tenth larger than the others, subquadrate, eleventh large, truncate at base, ovate, acuminate; prothorax narrower than the head, cordate, and having two lateral sulci and a transverse one, angular in the middle, base impressed on each side; elytra a little longer than broad, slightly rounded laterally, shoulders oblique, somewhat prominent, base trifoveate; abdomen a little shorter and narrower than the elytra; first dorsal segment foveate on each side but neither deeply nor broadly; legs strong, tibiæ slightly incurved, the intermediate ones have a minute apical spur; metasternum deeply depressed, last ventral segment strongly impressed. Male. Length 1·80–2 mm.

Several males have been captured near Salisbury, but no females; the latter probably remain at the roots of grass, while the former were caught flying at sunset in their search for a mate.

GEN. BATRISODES, Reitter,
Vehr. Naturf. Ver. Brünn., xx., p. 205.

Body oblong, head large, quadrate, eyes large and situated behind in an angle, last joint of palpi fusiform; antennæ moderately elongate, club most often more or less triarticulate; prothorax cordate, trisulcate lengthways but sometimes bisulcate ; elytra more or less rounded at the shoulders, or oblique, sutural stria entire, dorsal one shortened ; abdomen a little narrower than the elytra, immarginate, obtuse and abruptly declivous at apex, first dorsal segment by far the largest, the others being hardly conspicuous at first sight.

This genus, which was formerly included in *Batrisus*, is a very distinct one, owing to the large size of the head, the position of the eyes, and the first dorsal segment very much larger than all the others put together.

Numerous in the Indo-Malayan region ; few only are recorded from Africa (West Coast, Gaboon, Abyssinia, East Coast).

BATRISODES NATALENSIS,
Plate XVI., fig. 20.

Oblong, rufous, shining, covered with a short pallid pubescence ; head large, quadrate, sinuate laterally, frontal part deflexed, slightly raised on each side above the insertion of the antennæ, transversely sulcate in front, and having two large foveæ nearly before the eyes, which are large ; antennæ elongate, third to eighth joints suboblong, seventh a little longer, ninth to tenth larger, ovate, eleventh larger, ovate, acuminate ; prothorax broader than the head, strongly cordate, rounded laterally, trisulcate longitudinally and unisulcate transversely, base bifoveate ; elytra large, slightly rounded laterally, subconvex, shoulders oblique and obtusely dentate, dorsal stria nearly straight, abbreviated before reaching the apex ; abdomen a little shorter than the elytra, attenuate at apex, first dorsal segment large, bi-impressed at base, median impression larger than the others, transverse, broadly and deeply excavate at apex, the anterior margin of the excavation is deeply bisinuate, the apical one bituberculate, the bottom has a median transverse laminiform process with a tubercle above it in the median part ; metasternum hardly sulcate ; legs elongate, all the femora thick, slightly sinuate behind, all tibiæ nearly straight. Male. Length 1·20 mm.

Hab. Natal (Frere). Three examples.

TRIBE BRYAXININI.

Body generally short and more or less globose and convex ; antennæ eleven-jointed, separate at base, head generally flat, trapezoid, bearing no antennal tubercles ; the maxillary palpi are well developed, but never of large size, the first one concealed, as usual, in the mouth, the second one more or less elongate, curved and clavate at tip, the third small, globular, and triangular, fourth ovate and acuminate—these organs vary to a certain extent; the abdomen is nearly always more or less marginate, the hind coxæ far apart, and the first ventral segment concealed under the metasternum; the tarsi have one single claw; the under part of the head bears nearly always a large, more or less obtuse, longitudinal carina.

This tribe is widely distributed throughout the world, but it has, comparatively speaking, few representatives in South Africa.

Gen. RYBAXIS, Saulcy,
Bullet. Metz., xiv., 1876, p. 96.

Body short, convex; antennæ eleven-jointed; maxillary palpi of moderate size, second joint clavate at tip, third globose, minute, fourth briefly oblong, obtusely acuminate; prothorax cordate and with three foveæ connected by a transverse sulcus; elytra with a sutural and a dorsal stria, the latter more or less abbreviate, lateral margin sulcate inwardly; abdomen marginate.

Owing to the epipleural sulcus of the elytra and the transverse sulcus of the prothorax, this genus has been removed from *Bryaxis*.

Rybaxis circumflexa, Raffr.
Bryaxis circumflexa, Raffr., Rev. d'Entom, 1882, p. 24.

Little convex, chestnut red, shining, covered with a very short pubescence; head rather elongate with the sides nearly straight, slightly attenuate in front, trifoveate, anterior fovea not so deep nor as well defined as the others; antennæ elongate, slender, third to seventh joints subelongate, decreasing in length, eighth nearly quadrate, club triarticulate; prothorax much broader than the head, nearly as long as broad, subcordate, lateral foveæ large, median one small, transverse sulcus angular; elytra a little longer than broad, slightly attenuate at base, shoulders oblique, noticeable, bifoveate at the base with the discoidal stria more or less abbreviate in the apical part, nearly straight or slightly sinuate at apex; first abdominal segment carrying two carinulæ diverging much, and more or less apart; legs strong, femora thickened, tibiæ hardly arcuate.

Male: Antennæ long, joints much elongate, ninth to tenth subcylindrical, slightly conical, more than twice as long as broad, eleventh oblong, much elongated; third ventral segment with a median, minute, compressed tubercle; last segment hardly impressed; intermediate trochanters obtusely dentate inwardly at base; apical margin of the elytra bisinuate.

Female: Antennæ shorter than in the male, joints ninth and tenth smaller, truncate in an obconical form, eleventh oblong and a little shorter; apical margin of the elytra truncate in a nearly straight line. Length 2·30–2·50 mm.

This species is apt to vary somewhat; the discoidal stria of the elytra is more or less abbreviate at tip; when longer than usual it is slightly arcuate at the end; when short it is straight; the carinulæ of the abdomen are always oblique and diverging, but they are more or less apart from each other; the size of the body varies very much. I have some examples from Abyssinia which are only 1·90 mm. in length, while others reach 2·80 mm.; I have not noticed so much difference in size among the South African examples.

This species seems to have a wide distribution in Africa.

Hab. Natal (Frere), Zambezia (Salisbury), Abyssinia.

Gen. BRYAXIS, Leach,
Zool. Miscell., iii., 1817, p. 85.

This genus is nearly similar to the preceding one, but the body is generally more parallel, the foveæ on the prothorax are of equal size and free, as there is no transverse sulcus, and the lateral margin of the elytra are not sulcate.

Bryaxis seems to be confined to Europe and the African and Asiatic Mediterranean shores (palæarctic fauna) and North America. Its occurrence in South Africa is doubtful, the only two specimens as yet recorded from this locality being two examples which I have obtained from Mr. Boucard, and reputed to have come from South Africa; these two specimens are females of a well-known European species (*B. hæmatica*).

It must be said, on the other hand, that this identical species is found also in North America, where it is, however, so rare that it is unknown to the American entomologists. I have in my collection Dejean's type of *B. obscura*, Dej., from North America, which is completely identical with *B. hæmatica*. This species is not rare in all the temperate parts of Europe, and I have discovered in Algiers a slight variety.

That the two specimens alluded to are females of *B. hæmatica* seems to me indubitable, but we must not necessarily infer that the case is proven, because in many cases the females of different species of *Bryaxis* are so similar that they are almost indistinguishable, while the males of these same species are very different. If the male is ever discovered in South Africa it may prove to be a different species, or perhaps a mere variety, such varieties being already known (var. *perforata*, Aubé; *tubericentris*, Raffr.).

BRYAXIS HÆMATICA, Reichenb.,
Monogr., p. 52, pl. ii., fig. 12.

Entirely rufo-ferruginous, covered with a rufous pubescence; head short, trifoveate, eyes large; antennæ rather thick, third to seventh joints briefly oblong, decreasing in length, eighth quadrate, ninth and tenth larger than the preceding ones and increasing, quadrate, transverse, eleventh large, ovate, acuminate; prothorax a little broader than the head and eyes, more attenuate in front than behind, and having three large, disconnected foveæ, the median one of which is a little smaller than the other two; elytra subquadrate, slightly attenuate at base, shoulders little noticeable, base bifoveate, dorsal stria straight but not produced beyond the third of the length; first dorsal segment large and with two nearly straight and sub-parallel, well-defined carinulæ reaching lengthways the third part of the disk and inclosing nearly one-third of the width; legs robust. Female. Length 1·70 mm.

Hab. South Africa (?).

SUB-GEN. REICHENBACHIA, Leach,
Vigor's Zoolog. Journ., vol. ii., 1826, p. 451.

This subdivision of the genus *Bryaxis* differs only from it by the more convex, globular form of the body and the very small median fovea in the prothorax.

Taken in themselves the characteristics above mentioned cannot be said to be generic, but it has been found advisable to divide the old genus *Bryaxis*, which comprises now several hundred species, into groups, and to give each group a name in order to facilitate the study.

It is always difficult, and in some instances nearly impossible, to identify with certainty an isolated female, and this difficulty holds good with the South African *Reichenbachia*.

The country in the world where this sub-genus is most abundant is Tropical America, then comes the Indo-Malayan region, the palæ-arctic fauna, and South Africa. Australia has no representative.

REICHENBACHIA SULCICORNIS, Raffr.,
Annal. Soc. Entom. Franc., 1895, p. 389.

Oblong, thick, ferruginous, covered with a subflexible pubescence; legs and antennæ rufous, the latter infuscate at tip; head longer than broad, trifoveate, frontal fovea larger than the others; antennæ strong, third to sixth joints oblong, sixth a little shorter, seventh quadrate, eighth slightly transverse, ninth to tenth a little larger than the preceding, and increasing, transverse, eleventh large; prothorax rather broad, lateral foveæ distant from the margin, the central one very minute; elytra subquadrate, slightly attenuate at base, trifoveate, dorsal stria nearly straight, abbreviate before the apex; first dorsal segment of the abdomen with carinulæ variable in length, always diverging and inclosing at the base one-fourth of the width of the disk; metasternum moderately raised and depressed.

Male: Last joint of antennæ larger than the others, oblong, and having underneath a strong, sulciform fovea; intermediate tibiæ with an apical spur, last ventral segment broadly but obsoletely impressed.

Female: Last joint of antennæ ovate. Length 1·50–1·60 mm.

Allied to *R. picticornis*, Reitt.; in the male the antennal club is very different; it is difficult to distinguish the female from that of the last-named species, but the penultimate joint of the antennæ is decidedly more transverse, and the last one shorter and stouter.

Hab. Bechuanaland (Vryburg), Zambezia (Salisbury).

REICHENBACHIA PICTICORNIS, Reitt.,
Deutsch. Entom. Zeit., 1882, p. 188, pl. ix., fig. 6.

Oblong, dark ferruginous, hardly pubescent, legs and antennæ light rufous, the latter, principally in the male, dark at tip; head subquadrate, trifoveate; antennæ of moderate size; third to sixth joints oblong, subcylindrical, seventh much shorter, nearly quadrate, the others variable in each sex; prothorax nearly as broad as long, broader than the head, more attenuate in front than behind, lateral foveæ of moderate size, distant from the margins, the median one minute and suboblong; elytra subquadrate, slightly attenuate at base and trifoveate, shoulders hardly conspicuous, dorsal stria abbreviate before the apex, slightly arcuate and curving slightly outward at tip; first dorsal segment of abdomen with little conspicuous carinula, shorter than half the length of the discoidal part, diverging and covering a quarter of the width; posterior tibiæ slightly incurved; metasternum rather convex.

Male: Antennal club large, piceous, covered with a denser greyish pubescence, eighth joint broader and shorter than the preceding one, transverse, ninth a little broader than the eighth and more than twice as long, transverse, tenth longer, subquadrate, eleventh oblong, truncate at base, acuminate at tip; intermediate tibiæ with an obtuse, hardly discernible spur at tip, last ventral segment depressed transversely; metasternum abruptly impressed at apex.

Female: Antennal club much smaller than in the male, ultimate joint hardly infuscate, eighth joint quadrate, ninth a little larger, quadrate, tenth larger than the preceding one, quadrate, eleventh ovate, hardly thickened, truncate at base and acuminate at tip. Female : Variety β. (?) Antennæ more slender than in type, piceous at tip, and with the eighth, ninth, and tenth joints slightly transverse. Length 1·50-1·60 mm.

Hab. Natal (Frere), Zambezia (Salisbury), and also the West Coast of Africa.

In the South African examples the colour is darker, and the last joint of the antennæ entirely dark, whilst in the West Coast specimens the ultimate joint is dark at the base only, and in the female the antennæ are entirely of the same colour ; this is, however, only a local variety.

The diagnosis of the female var. β is made from one example ; it is impossible to decide if it belongs to a distinct species or not. The habitat of that example is Zambesia (Salisbury).

REICHENBACHIA DECIPIENS.

Closely allied to both *R. sulcicornis* and *R. picticornis*, but the only example I have seen being a male, proves to be a very distinct species. The colour is a little lighter, the antennæ more slender, the third to sixth joints long, subcylindrical, seventh much shorter than the sixth, but still longer than broad, eighth quadrate, ninth a little longer than eighth, quadrate, tenth larger and very slightly longer than broad, eleventh fusiform, rather elongate and slender ; the carinulæ of the abdomen are more definite and less divergent ; the intermediate trochanters have a very small acute tubercule at base ; the metasternum is broadly depressed, and the last ventral segment is broad and impressed. Male. Length 1·70 mm.

Hab. Zambezia (Salisbury).

REICHENBACHIA SUBPUBESCENS.

Oblong, not very thick, ferruginous, covered with a moderately dense and long greyish pubescence, under side of abdomen and antennæ infuscate at tip, legs rufous ; head longer than broad, hardly attenuate in the anterior part, trifoveate ; posterior foveæ transverse ; antennæ slender, joints third to sixth elongate, seventh much shorter, eighth quadrate, ninth hardly broader, but longer, tenth larger, briefly ovate, eleventh of moderate size, ovate, acuminate ; prothorax a little broader than the head, rounded laterally, more attenuate in the anterior than in the posterior part, lateral foveæ of moderate size, greatly distant from the margin, median one punctiform ; elytra moderately elongate, hardly attenuate at base, trifoveate, dorsal stria straight, much attenuate before the apex; abdominal carinulæ hardly divergent, reaching lengthways the median part of the disk, and across a fifth of the width; metasternum broadly impressed, anterior trochanters, and also the intermediate ones with a minute but acute tubercle at base; anterior tibiæ sinuate inwardly before the apex, posterior ones arcuate, last ventral segment with a minute tubercle at base and totally impressed. Male. Length 1·59 mm. One example.

Hab. Natal (Frere).

I have also seen two females, which I describe as possible varieties of the present species, as I do not deem it prudent to consider them as distinct species owing to the small number of examples examined, but I would not be surprised if the discovery of the male sex would prove them to belong to a distinct species.

Female: Var. β vel. spec.(?) Thicker, unicolor; antennæ entirely infuscate, joints ninth and tenth more quadrate, dorsal stria slightly arcuate inwardly. Length 1·40 mm. One example from Zambezia (Salisbury).

Female: Var. γ vel. spec.(?) Antennæ thicker, entirely rufous, dorsal stria slightly arcuate; carinulæ of the abdomen shorter and very near each other. Length 1·40 mm. One example from Zambezia (Salisbury).

Reichenbachia discreta.

Rather broad and thick, covered with a subflexible pubescence; legs rufo-testaceous, antennæ rufous, head, prothorax and elytra very slightly punctulate; head longer than broad, not attenuate in front, deeply and equally trifoveate; antennæ moderately elongate and slender, joints seventh to tenth elongate, decreasing in length, eighth subquadrate, but nevertheless a little longer than broad, ninth hardly larger, trapezoidal, tenth more trapezoidal, eleventh ovate, strongly acuminate; prothorax broader and shorter than the head, sides rounded, lateral foveæ of moderate size, distant from the suture, the median one much smaller than the others; elytra rather broad and slightly depressed, a little attenuate at base and bifoveate, shoulders not prominent, dorsal stria entire, arcuate and ending in the sutural angle; abdomen a little broader than the elytra, carinula slightly diverging and reaching to half the length of the disk and inclosing more than one-quarter of the width; metasternum hardly impressed; posterior tibiæ slightly incurved. Female. Length 1·80 mm.

Although I have seen only one female example, I do not hesitate in considering it as a very distinct species, differing in many points from the preceding ones.

Hab. Zambezia (Salisbury).

REICHENBACHIA AFRA.

Somewhat short and broad, ferruginous, covered with a subflexible pubescence, legs reddish; head a little longer than broad, and having three foveæ, the anterior of which is larger than the others and oblong; antennæ short, thick, variable in both sexes; prothorax broader than the head, not longer, lateral foveæ moderate, distant from the margin, median one smaller; elytra subquadrate, hardly attenuate at base and tripunctate, dorsal stria slightly arcuate, and shortened at some distance from the sutural angle.

Male: Third to sixth joints of antennæ briefly oblong, decreasing in length, seventh and eighth quadrate, ninth strongly transverse, slightly produced inwardly, tenth a little broader than the preceding one but twice as long and transverse, eleventh very large, subglobose, truncate at base and broadly but not deeply impressed and the impression very shining, obtusely acuminate at tip; abdominal carinæ parallel, and inclosing a third of the width of disk; metasternum obtusely gibbose laterally; intermediate tibiæ provided with a long strong ante-apical spur.

Female: Third to fifth joints of antennæ briefly oblong, sixth quadrate, seventh to eighth slightly transverse, ninth and tenth transverse, increasing in size, eleventh briefly ovate, truncate at base, oblique externally towards the apex and obtusely acuminate; carinula of abdomen slightly divergent, shorter than in the other species, and inclosing hardly the fourth part of the width of disk; metasternum simple. Length 1·50 mm.

This species may at once be distinguished from the others by the short body and thick antennæ.

Hab. Zambezia (Salisbury). One pair only.

REICHENBACHIA DIVERSA, Raffr.,
Rev. Entom., vi., 1887, p. 36.

Short, rather thick, piceous or piceous red, elytra sometimes of a lighter colour on disk, legs obscure red; head subquadrate, slightly attenuate in the anterior part, trifoveate, anterior fovea larger than the others and oblong; antennæ thick, third joint a little longer than the others, fourth and sixth subovate, others variable in both sexes; prothorax hardly broader than the head, lateral foveæ distant from the margin, the median one punctiform; elytra large, slightly attenuate at base and bifoveate, shoulders well defined, dorsal stria slightly incurved and abbreviate before the apex; carinulæ of abdomen short, more or less divergent and inclosing the third part of the width of disk; posterior tibiæ incurved, intermediate ones slightly sinuate; metasternum not depressed.

Male: Antennæ thicker than in the female, and more clavate, seventh joint of antennæ hardly transverse, eighth and ninth much transverse, increasing gradually and slightly produced inwardly, tenth much larger than the others, trapezoidal, slightly transverse, eleventh large, ovate, strongly dentate inwardly at base; intermediate tibiæ very briefly spurred at apex; metasternum and last ventral segment slightly depressed.

Female: Antennæ very much less clavate than in the male, seventh joint quadrate, eighth a little transverse, ninth and tenth increasing in size, trapezoidal, slightly transverse, eleventh ovate, obtusely acuminate at tip. Length 1·40–1·60 mm.

Hab. Cape Colony (neighbourhood of Cape Town, Stellenbosch).

Male: Var. β. Much smaller than the type form, 1·20 mm., completely rufous.

Hab. Neighbourhood of Cape Town (Mowbray).

REICHENBACHIA PERINGUEYI.

R. diversa var. *unicolor*, female, Raffr., Rev. Entom., vi., 1887, p. 36.

Oblong, piceous red, legs and antennæ rufous at base; antennæ of moderate size, third to sixth joints oblong, sixth ever so little shorter, seventh quadrate, eighth to tenth transverse, gradually increasing, eleventh ovate, slightly oblong; lateral foveæ of the prothorax distant from the margins and not much larger than the median one; elytra as long as broad, little attenuate at base and trifoveate, dorsal stria hardly arcuate, and stopping abruptly without reaching the apex; abdominal carinulæ extremely short, slightly diverging and inclosing a fifth of the width of the disk.

Male: Ninth and tenth joints of antennæ broader and more transverse than in the female, eleventh suboblong; intermediate tibiæ with an apical, rather strong spur, last ventral segment slightly depressed transversely. Length 1·30 mm.

This species is very closely allied to *R. diversa*; having only one female for my original description, I described it at first as possibly a variety of the afore-named species; I have since obtained the male, and it proves to be a very distinct species, smaller and more slender than the others.

Hab. Cape Colony (Stellenbosch).

REICHENBACHIA ACHILLIS, C. Schauf.,
Catalog. Tijdschr. Ent., xxxi., p. 20.
Bryaxis crassipes, Raffr., Rev. Entom., vi., 1887, p. 36.

Rather elongate, piceous black, elytra red, legs and antennæ brown, body with a hardly noticeable pubescence, and shining; the colour turns sometimes to red all over, principally in the female; head subquadrate, foveæ shallow, anterior one larger than the others; antennæ rather elongate, third to sixth joints elongate, seventh much shorter, eighth quadrate, ninth larger, quadrate, tenth trapezoidal and larger, eleventh large, ovate, acuminate; prothorax larger than the head, rather elongate, oblique at base, lateral foveæ placed close to the margin, of moderate size, median one minute, base punctulate; elytra somewhat elongate, slightly attenuate at base and trifoveate, shoulders little pronounced, dorsal stria nearly entire and slightly bisinuate; abdomen shorter than the elytra, carinulæ extremely short.

Male: Antennæ more slender and longer than in the female; anterior tibiæ slightly dilated in the middle inwardly and more or less minutely tuberculate before the apex, intermediate ones slightly sinuate inwardly and with a strong spur at tip, posterior ones arcuate, all the femora thick; metasternum strongly and broadly impressed, minutely bituberculate close to the intermediate coxæ; fourth and third ventral segments with a large, oblong fovea reaching also the tip of the second one, ultimate segment rugoso-punctate, having a deep subtriangular fovea at the base, and impressed transversely at apex. Length 1·80 mm.

Female: Antennæ a little shorter than in the male, legs not so thick; metasternum slightly impressed, last ventral segment rugoso-punctate, second dorsal one acutely mucronate in the middle of the apical part. Length 1·60 mm.

Hab. Cape Colony (Stellenbosch). Rare.

Female: Var. *bimucronata*. Second and third dorsal segments sharply spinose in the middle of the apical part; sometimes the first segment is very briefly spinose in the median part of the apex. Length 1·40–1·60 mm.

Hab. Cape Colony (Muizenberg, in the neighbourhood of Cape Town).

This female variety (*bimucronata*) is a very interesting one, because dimorphism is extremely rare among females, and I know of no such occurrence in the whole family. I have captured also the male of this variety, which does not show the slightest difference with the typical male captured with the typical female. It is also worthy of note that this variety does not occur with the type form, which seems so far to be restricted to Stellenbosch, while the former has only been met with at Muizenberg, where I found several females and one male at the foot of grass; but on the slopes of Table Mountain I have met with some female examples in which the little spine on the first dorsal segment has disappeared; and this is to a certain extent a transitory form between the type and the variety *bimucronata*.

Tribe GONIACERINI.

Body variable in shape; head elongate and provided with an antennal tubercle, depressed laterally in front and more or less dilated; eyes placed forward; antennæ geniculate; maxillary palpi of moderate size; abdomen marginate; first ventral segment of abdomen conspicuous, and projecting over the posterior coxæ, posterior and intermediate coxæ apart from each other; one tarsal claw provided with a setiform appendage.

In this tribe the antennæ are geniculate as in the *Curculionidæ*, and not unlike therefore *Metopius* from America, but they are greatly differentiated by other characteristics.

The insects included in this tribe are not numerous, and seem to be exceedingly rare; they occur in America and Africa, but are much more numerous in America, and four species included in two genera were so far known from Abyssinia, Zanzibar, and the Gaboon; but my friend Mons. Eugène Simon, the well-known arachnologist, who visited South Africa in 1892, has discovered a fifth one in the Transvaal. These five species are known from single specimens only, three of which I have myself captured in Abyssinia and Zanzibar flying at sunset.

Gen. OGMOCERUS, Raffr.,
Rev. Entom., vol. iii., 1882, p. 7.

Body oblong, somewhat depressed; head provided with a frontal tubercle, depressed laterally in front; eyes set forward; antennæ geniculate, elongate, eleven-jointed, club triarticulate; maxillary palpi moderate, triarticulate, first joint rather elongate, slightly incurved, thickened at tip, second smaller, third briefly fusiform; abdomen broadly marginate; legs somewhat elongate; first joint of antennæ minute, second and third subcylindrical, third longer than the others.

Some species included in the genus (*O. giganteus, agymsibanus*) are among the largest known *Pselaphidæ*.

OGMOCERUS RUGOSUS, Raffr.,
Ann. Soc. Ent. Franc., 1895, p. 390.

Somewhat elongate and depressed, ferruginous, slightly covered with a yellowish pubescence, entirely rugoso-punctate, more so on he head and prothorax; head subparallel laterally, and having on the vertex two tomentose foveæ, antennal tubercle moderately broad and depressed, and having a tomentose fovea on the summit; antennæ rather thick, geniculate, first joint very long, one-half the length of the antennæ, sinuate, second one globose transversely, third to sixth transverse, seventh a little larger, transverse, globose, eighth to tenth transverse, increasing in width, eleventh very briefly ovate; prothorax slightly cordate, having a longitudinal sulcus and a large fovea past the median part, transverse sulcus obsolete; elytra subquadrate, subdepressed, with the sides nearly parallel, base bifoveate, dorsal stria abbreviate past the median part; abdomen longer than the elytra, subconvex; legs robust; metasternum hardly impressed. Female. Length 3·50 mm.

This species is smaller than the others, the punctures are really small, tubercles larger on the head and prothorax and smaller on the elytra and abdomen.

Hab. Transvaal (Hamman's Kraal near Pretoria).

Tribe PSELAPHINI.

Head long and narrow; antennæ clavate; maxillary palpi generally unusually long; prothorax more or less ovate; elytra more or less triangular; abdomen with a broad margin, first segment very developed behind the hind coxæ, but hidden from view by a pale, glandular pubescence; tarsi always with a single claw.

This tribe is well represented in Europe and Australia, but has only a few representatives in Africa.

Gen. PSELAPHUS, Herbst,
Käf., iv., 1792, p. 106.

Elongate, attenuate in front; head elongate, more or less sulcate; maxillary palpi very long and slender, first joint filiform and shorter than the second, which is also filiform, but clavate at tip; third minute, subglobose or triangular, fourth long, filiform, strongly clavate at tip; antennæ elongate, club triarticulate; prothorax oblong; elytra much attenuate at base, ampliate behind; first dorsal abdominal segment large; legs elongate, slender, first tarsal joint minute, second clavate at tip, third cylindrical, a single claw.

The genus is represented in every part of the world.

PSELAPHUS LONGICEPS, Raffr.,
Rev. Entom., vi., 1887, p. 33.

Chestnut red, very shining, smooth, briefly nigro-setose here and there on the prothorax and elytra, the setæ more numerous on the abdomen; legs paler; palpi testaceous; head strongly elongate, four times longer than broad, subparallel laterally, neither broadly nor deeply sulcate in front, abruptly raised close to the eyes, and obsoletely bispinose; palpi much elongate, and with a few whitish setæ, last joint slightly sinuate, neither strongly nor abruptly clavate, apex of the clava sulcate; antennæ reaching further than the base of the elytra, first joint elongate, subcylindrical, closely punctured, second subquadrate, a little longer than broad, third longer, oblong, fourth to eighth shorter, obconical, ninth to tenth much larger, obovate, ultimate one large, obovate, acuminate; prothorax hardly shorter than the head, a little broader, attenuate in front and behind, very slightly sinuate on each side before the base, and without any sulcus or fovea; elytra once and a half longer than the prothorax, much broader, much attenuate at base; shoulders oblique, well defined, sides slightly rounded, covered at apex with long, black setæ, as well as with a glandular ochreous pubescence, sutural stria subcarinate, entire, the dorsal one close to the suture; abdomen hardly broader than the elytra and a little shorter; metasternum and abdomen simple underneath. Female. Length 1·40 mm.

Allied to *P. filipalpis*, Reitt., from the Gold Coast, but differs by its larger size and more slender form.

Hab. Natal (Frere); one example. Occurs also in Zanzibar.

Gen. PSELAPHISCHNUS.

Not much elongate, attenuate in front; head elongate; maxillary palpi much elongate, first joint filiform, short, second filiform, clavate at tip, third minute, subtriangular, fourth large, globose at base, from there filiform and subulate for a great length; antennæ thickened, joints transverse; abdomen broad and broadly marginate, first dorsal segment larger than the others; legs thick and short.

Resembles much *Pselaphus*, but is much shorter and broader, the margin of the abdomen is also broader, the antennæ are thick, and the maxillary palpi very different, the last joint instead of being clavate at the apex is clavate at the base and thin and sharp at tip.

PSELAPHISCHNUS SQUAMOSUS,
Plate XVII., fig. 2.

Chestnut brown, totally covered with ochreous squamæ; palpi

testaceous, tarsi red ; head slightly attenuate in front, sulcate longitudinally, and having two obsolete foveæ set in the middle between the eyes, vertex slightly convex, cheeks glanduloso-squamose behind ; antennæ thick, first joint large, slightly obconical, second a little narrower, quadrato-transverse, third to eighth smaller than the others, equal to each other and transverse, ninth a little larger, transverse, eleventh large, much truncate at base, obtuse at apex ; prothorax ovate, convex, a little broader than the head, sinuate at base, truncate, two lateral foveæ and a median minute one ; elytra triangular, not broader at base than the prothorax, oblique laterally, more than three times broader at apex, apical margin with very pale glandulose squammæ ; abdomen longer than the elytra, slightly rounded laterally and a little broader, very broadly marginate, rounded at apex, first dorsal segment larger than the others, and with the median apical part having an acute, straight tubercle ; metasternum convex, first ventral segment covered with a whitish glandulose pubescence, second one large ; abdomen slightly concave underneath ; legs slightly compressed. Male.

This very fine insect, owing to the squamose pubescence and short and triangular elytra, is not unlike *Pselaphus opacus*, Schauf., from the Amazon.

Hab. Cape Colony (Cape Town and neighbourhood).

Tribe CTENISTINI.

Head provided with an antennal tubercle, cheeks more or less dilated in front in a tubercle ; maxillary palpi conspicuous, most often penicillate ; first ventral segment of abdomen hidden by the posterior coxæ ; claws double and equal in size ; the pubescence is always squamose.

This tribe differs from the *Pselaphini* by the mucronate sides of the head in the anterior part, the first ventral segment of the abdomen is short, and more or less covered by the hind coxæ, the two claws of the tarsi are equal, and the pubescence is always squamose.

This tribe is found everywhere. The insects belonging to it are generally met with in damp or swampy places, and may be caught flying at sunset ; some are found under stones, others in ants' nests.

Gen. LAPHIDIODERUS, Raffr.,
Rev. Entom., 1887, p. 20.

Subelongate, subdepressed ; head provided with a strong antennal tubercle ; epistoma with the sides angular and dilated ; antennæ approximate at base, thick, clavate at apex, similar in each sex; maxillary palpi large, first joint inconspicuous, second subelongate, arcuate, clavate at tip and provided with an apophyse, third oblong, angular externally and with an apophyse, fourth oblong, dilated at base, provided with an apophyse and strongly acuminate at tip; prothorax subpentagonal, tuberculate and foveate; elytra short, dorsal and sutural striæ entire; abdomen large, broadly marginate; legs long, femora clavate; tibiæ slender; posterior coxæ very much separated; intermediate and posterior trochanters elongated, clavate; tarsi subelongate and having two equal claws; head armed underneath with a long, strong, recurved spine under the eye.

Very closely allied to the African genus, *Desimia*, Reitt., and more especially to the American one, *Ctenisis*, Raffr., but differs from both by the antennæ having a three-articulated club in both sexes, instead of the long cylindrical four-jointed club of the male in *Desimia* and *Ctenisis*. It differs also from *Ctenisis* by the shape of the antennæ and that of the palpi, the third and fourth joints of which are not transverse.

This genus is purely a South African one, and includes only two species closely allied to each other.

LAPHIDIODERUS CAPENSIS, Raffr.,
Rev. Entom., 1887, p. 21, pl. i., figs. 2, 3.

Piceous red or obscurely rufous, shining, covered all over with sparse ochreous squamæ, all the foveæ and the sutural part with yellowish glandulose squamæ; head subelongate, strongly depressed, narrowed in the anterior more than in the posterior part; antennal tubercle large, subdivided, minutely foveate at base, two large foveæ between the eyes, which are large; antennæ reaching the median part of the elytra, gradually clavate, all the joints as broad as long, penultimate ones increasing, eleventh oblongo-ovate, obtuse at tip; prothorax longer than broad, attenuate in front with the sides not quite straight, and having three longitudinal deep sulci reaching further than the median part, base depressed transversely; elytra not longer than the prothorax, attenuate at base, little oblique laterally, deeply bifoveate at base, sutural and dorsal striæ strong and entire; abdomen once and a half longer than the elytra, first dorsal segment a little shorter than the following one; metasternum deeply sulcate; anterior tibiæ much arcuate.

Male: Antennæ with the eighth to tenth joints increasing in size, eleventh oblong, club quadri-articulate; metasternum having on each side a large subconical tubercle acute at tip.

Female: Eighth joint of antennæ a little shorter than the preceding one, ninth and tenth increasing in size, little elongate, eleventh oblongo-ovate: metasternum having on each side an ovate depressed tubercle not acute at tip. Length 1·60–1·80 mm.

Mr. Péringuey discovered this species near Cape Town in the galleries of an ant (*Bothroponera pumicata*). I have found it in nearly the same locality under stones, and there were no ants; it is abundant in June and July on the slopes of the Lion's Rump in Cape Town.

LAPHIDIODERUS BREVIPENNIS.

Closely allied to the preceding species, but the head is more elongate; antennæ shorter, eighth to tenth joints a little longer than broad, slightly transverse, eleventh oblong, acuminate at tip; prothorax shorter than broad, sulci shorter, not reaching the median part; elytra almost shorter than the prothorax, much transverse, attenuate at base, slightly rounded laterally, dorsal stria arcuate; abdomen twice as long as the elytra; metasternum sulcate.

Male: Last joint of antennæ hardly longer than the others, club triarticulate; metasternum having on each side an ovate tubercle ending in a minute tooth.

Female: Metasternum having on each side an ovate tubercle not sharp at tip. Length 2 mm.

Hab. Cape Colony (environs of Cape Town, Mowbray); two examples.

This species is so very much like the preceding one that a comparative diagnosis may prove useful.

Capensis.	*Brevipennis.*
All the joints of the antennæ longer than broad, antennal club of male four-jointed.	Joints eighth, ninth, and tenth transverse; antennal club of male trijointed.
Prothorax longer than broad, basal sulcus longer than half the length.	Prothorax as broad as long, basal sulcus shorter than the prothorax by half the length.
Elytra short, transverse, longer than the second dorsal segment of the abdomen.	Elytra very short, very transverse, not longer than the second dorsal segment of the abdomen.
Abdomen once and a half as long as the elytra.	Abdomen twice the length of the elytra.
Metasternum of the male with two conical and acute tubercles.	Metasternum of male with two flat tubercles bearing a small tooth behind.

Gen. CTENISTES, Reichenb.,
Monogr. Pselaph., 1816, p. 75.

Head elongate, provided with an antennal tubercle, cheeks obtusely tuberculate in front; antennæ of the male with the third to seventh joints minute, moniliform, eighth to eleventh cylindrical and forming a large elongate club, those of the female sensibly increased towards the apex and the club triarticulate; maxillary palpi large, the last three joints provided with an apophyse, and the third and fourth very transverse; prothorax more or less obconical; elytra moderately elongate; abdomen marginate; legs rather elongate.

The genus is represented in every part of the world.

Ctenistes australis, Raffr.,
Rev. Entom., 1887, p. 25.

Fulvous, elytra, legs, and palpi paler, body covered with ochreous squamæ; head oblong, trifoveate; antennal fovea much smaller than the others, oblong; eyes very large; palpi short, second joint slightly arcuate and a little clavate at apex, third and fourth transverse, ovate, appendages short; prothorax a little longer than broad and subconical, hardly rounded laterally, briefly foveate at base; elytra little elongate and little attenuate at base, sides slightly rounded, shoulders oblique, dorsal stria entire, slightly arcuate; second dorsal segment of the abdomen twice as long as the first. Length 1·90–2·20 mm.

Male: More slender, smaller and paler than the female; first and second joints of the antennæ larger than the others, third to seventh very minute, moniliform, eighth equal in length to the seven preceding ones, ninth nearly shorter by one-half, tenth a little longer than the ninth, eleventh equal to the preceding and acuminate.

Female: Thicker, larger and darker than the male; antennæ thick, nearly one-third shorter than in the male, the first two joints larger than the following ones, third longer, oblong, fourth to sixth almost of equal size and slightly shorter than the third, seventh nearly twice as long as the preceding one, eighth much smaller than the sixth, the three apical ones the largest of all, clavate, ninth subglobose, tenth subquadrate, eleventh nearly twice as long as the tenth, oblong, acuminate at tip.

In this species the palpi are smaller than usual and the last two joints ovate and transverse; it resembles C. zanzibaricus, Raffr., but it is smaller and lighter in colour; the female is very unlike the male being larger and darker.

Hab. Cape Colony (Stellenbosch, Paarl), Natal (Frere, Escourt).

CTENISTES IMITATOR, Reitter,
Deutsch. Ent. Zeit., xxvi., 1882, p. 179.

Fulvous, elytra paler, palpi testaceous, body covered with ochreous squamæ; head elongate, trifoveate, anterior fovea a little smaller than the others; antennal tubercle hardly sulcate; eyes large; palpi long, second joint arcuate at apex and clavate, third subequal in length and width and produced in a very angular shape outwardly, fourth fusiform, transverse, appendages long; prothorax a little longer than broad, slightly conical, hardly rounded laterally, base with oblong foveæ; elytra rather elongate, oblique laterally, attenuate at base, oblique at shoulders, dorsal stria entire, hardly arcuate; abdomen with the dorsal segment hardly twice the length of the first.

Female: Antennæ slender, elongate, third to sixth joints elongate, slightly decreasing in length, seventh longer than the third, eighth subglobose, ninth ovate, nearly twice the thickness of the seventh but nearly equal in length, tenth hardly thicker than the preceding one but longer, eleventh equal in length to the two preceding ones put together but hardly thicker, subcylindrical, acuminate. Length 2·26 mm.

This species differs from the preceding one by the more elongate head, the much more slender antennæ, the elytra more narrowed at base and more arcuate laterally.

I cannot detect any appreciable difference between a Natal example and Reitter's type from the Gold Coast, except that it is more densely squamose and that the antennæ are very slightly longer, but in the absence of the male it is not possible to consider it as a distinct species.

Hab. Natal (Frere); one female. Originally described from Western Africa (Gold Coast).

GEN. ODONTALGUS, Raffr.,
Rev. Mag. Zool., 1877, p. 8, pl. iii., fig. 5.

Short, thick, convex; head provided with an antennal tubercle, cheeks hardly tuberculate in front, a palpal fovea underneath; eyes more or less conspicuously divided by a squamose canthus; first joint of maxillary palpi inconspicuous, second large, elongate, strongly clavate, third smaller, subtriangular, fourth large, elongate and very much clavate; antennæ elongate or thick, club of the male triarticulate, that of the female biarticulate, eighth joint smaller than the others; prothorax subconical, tuberculate; elytra much broader than the prothorax, carinate; abdomen marginate; intermediate coxæ little distant, posterior ones very far apart; metasternum large; all the trochanters elongate; tarsi with two claws equal in length.

The general facies is very different from that of *Ctenistes*; it is much shorter and thicker, and the shape of the palpi is entirely different.

The geographical distribution is somewhat peculiar. I first discovered two species in Abyssinia (one of which has been also found since in Natal), another one inhabits West Africa (Gold Coast), a fourth one occurs in Zanzibar, but the genus is also met with in Sumatra and Borneo.

ODONTALGUS VESPERTINUS, Raffr.,
Rev. and Mag. Zool., 1877, p. 9.

Short, thick, chestnut brown or piceous, body covered with greyish squamæ, palpi and tarsi rufo-testaceous, foveæ and sutures filled with glandular whitish squamæ; head depressed, strongly trifoveate, antennal tubercle slightly sulcate; antennæ rather thick, first joint large, subelongate, second larger than the following ones, subcylindrical, third slightly obconical, equal in length to the preceding one but more slender, fourth to seventh shorter and decreasing a little in length, eighth minute, very transverse, the other differing in each sex; prothorax slightly conical, sinuate laterally, with several impressions and having four tubercles arranged in the shape of a cross, the three anterior ones are the largest and oblong, the posterior one is very much smaller and rounded; elytra broader than the prothorax, subquadrate with the sides slightly rounded and ampliate beyond the median part, hardly attenuate at base, shoulders oblique, well developed, base bifoveate and broadly although obtusely tricarinate lengthways (suture included); abdomen abrupt behind, first dorsal segment a little larger than the others, and the basal three slightly and bluntly carinate in the middle of the apical part; anterior trochanters tuberculate at tip.

Male: Antennal club triarticulate, ninth joint subquadrate with the angles rounded, tenth similar but a little larger, eleventh broader and more than twice the length of the preceding one, subcylindrical, rounded at tip; metasternum very deeply excavate all over, bottom of excavation foveate in the anterior part, sulcate in the posterior and filled with glandulose squamæ, sides with an erect dentiform tubercle before the median part; whole of abdomen deeply concave underneath, second segment foveate in the middle, apical one dentate on each side; last dorsal segment obtusely carinate in the middle and slightly impressed on each side.

Female: Club of antennæ biarticulate, ninth joint hardly larger than the seventh, transverse, tenth much larger, slightly transverse, eleventh large, ovate; metasternum raised and plane, very much sulcate lengthways in the middle, glanduloso-squamose, and with another slender transverse anterior sulcus; abdomen slightly convex underneath, last dorsal segment depressed. Length 1·50–1·60 mm.

I cannot detect any difference between the South African and the Abyssinian forms, except that the latter is a trifle larger (1·70 mm.). The striking sexual characters of the metasternum and abdomen are exactly the same.

Hab. Natal (Frere). Very abundant in Abyssinia.

ODONTALGUS COSTATUS,
Plate XVII., fig. 9.

Stout, much attenuate in front, chestnut brown, totally covered with ochreous squamæ and set with large, shallow, sparse and squamæ-bearing punctures, palpi and tarsi testaceous; head longer than broad, sinuate laterally, vertex subgibbose and obsoletely trisulcate, frontal part with a large median foveæ; eyes large, divided by a squamose canthus; antennæ strong and thick, first joint large, second slightly transverse, third to ninth very transverse, tenth larger, eleventh large, ovate, obtuse at tip; palpi a little short; prothorax narrower than the head, eyes included, irregular and plurigibbose, much constricted laterally in front and having five foveæ; elytra shorter than the prothorax and twice as broad, rounded at the shoulders, sinuate laterally, obtusely but strongly tricostate, external costa abbreviate a little before the median part, intermediate one entire, oblique towards the shoulders and sinuate from there, internal one entire, slightly arcuate, suture hardly raised, bifoveate at base; abdomen stouter than the elytra, first dorsal segment depressed at base and more densely squamose, tricostate and bituberculate at apex, second and third with five tubercles, median tubercle distant from the apex and larger than the others, fourth segment strongly unituberculate in the middle, fifth inflexed and simple underneath; metasternum smooth and shining, delicately sulcate longitudinally and having close to the intermediate coxæ a large, squamose fovea, and on each side a large, triangular compressed tooth; second and third ventral segments moderately smooth in the middle and shining, the fifth one deeply emarginate in an arcuate form; femora thick, sulcate inwardly, anterior and intermediate tibiæ hardly sinuate while the posterior ones are strongly so. Female. Length 1·60 mm.

The facies of this insect is very different from that of the other *Odontalgus*; the antennæ are thick, and the joints transverse, the tubercles on the prothorax are very prominent; the carinæ on the elytra are much stronger, and there is no trace of lateral stria; the segments of the abdomen are provided with tubercles, which are absent in the other species. At first sight one feels inclined to consider this insect as belonging to a genus distinct from *Odontalgus*, but a close inspection reveals no generic difference.

Hab. Cape Colony (Cape Town, slopes of Table Mountain and Lion's Rump). Apparently very rare.

TRIBE TYRINI.

Body variable, mostly always attenuate at tip; head provided with a more or less conspicuous antennal tubercle, cheeks simple laterally in the anterior part; first ventral segment hidden by the posterior coxæ or the metasternum; abdomen most often marginate; intermediate trochanters always elongate, the others often so; claws of tarsi double, equal in length or subequal; pubescence always simple, mostly always elongate, sometimes short, but never squamose.

This tribe is closely allied to the preceding one, and differs chiefly by the hair-like pubescence and the non-prominent sides on the frontal part of the head.

The tarsi have two equal claws generally, but in some few insects from the Malayan Archipelago and Australia these claws are unequal.

The tribe includes most of the finest and largest *Pselaphidæ*, and is better represented in Africa than anywhere else, but it is also pretty numerous in Asia and Australia. One of the most important genera, *Centrophthalmus*, more numerous in Asia but extending also in Africa to the East and West Coasts, Abyssinia, and Algiers, is not found in South Africa.

GEN. TMESIPHORUS, Le Conte,
Bost. Journ. Nat. Hist., vi., 1850, p. 75.

SINTECTES, Westw.

Oblong, head elongate and with an ocular canthus underneath more or less produced and spinose; maxillary palpi strong, first joint inconspicuous, second arcuate, clavate and appendiculate, third more or less oblong, angular in the median part outwardly and appendiculate, fourth dilated externally, acute at tip; antennæ strong, club triarticulate; prothorax more or less cordate; elytra more or less briefly costate; abdomen marginate, bi- or tri-carinate, and sometimes without carinæ at all; posterior coxæ apart, tarsi rather elongate, third joint a little longer than the second, claws double, equal; pubescence variable, long or extremely short.

Species of this genus occur in North America, Australia, Asia, and

Africa, but they vary considerably; most of the Asiatic species have a long, soft pubescence, and have no abdominal carinæ, whilst the Australian, American, and African ones as well as few Asiatic have a very short, bristly pubescence, and two or three carinæ on the first dorsal segment of the abdomen, which impart to them a totally different appearance.

Few species are known from Africa, where they seem to be rare.

TMESIPHORUS RUGICOLLIS.

Oblong, piceous red, elytra chestnut red, and antennæ ferruginous or entirely ferruginous or rufous, palpi and tarsi testaceous, body covered with a pubescence having a somewhat golden tinge; head very closely rugoso-punctate, longer than broad, constricted in front of the antennal tubercle, which is ampliate and deeply divided, and having between the eyes two large foveæ, slightly transverse; between these the vertex is carinate vertically, and raised transversely behind, the infra-ocular canthus obtuse; third joint of palpi briefly oblong, angular outwardly in the middle, appendiculate, fourth broadly and roundly dilated; antennæ robust, first joint large, cylindrical, second quadrate, third to eighth a little smaller than the others, third to seventh slightly transverse, the eighth more so, club large, ninth and tenth trapezoidal, tenth a little larger and slightly transverse, eleventh truncate at base, ovate, obtuse at tip; prothorax closely rugoso-punctate, subcordate, subgibbose, rounded laterally in front, and from there sinuate on account of a large fovea situated on the side, having in the median part above the base an acute tubercle and an oblong, basal fovea, the bottom of which is smooth; elytra robust, sparsely punctured, subquadrate, slightly transverse, with the shoulders oblique, clavate, bifoveate at base and obtusely costate in the disk; this costa is abbreviate above the apex, and the suture costiform; abdomen larger than the elytra, slightly rounded laterally, somewhat convex, the two first segments tricarinate; legs strong, anterior tibiæ thickened in the middle and arcuate.

Male: Ninth joint of antennæ depressed inwardly, tenth totally and broadly excavate, dentate inwardly, and less so outwardly, eleventh ovate, neither deeply nor broadly impressed, more or less unidentate; metasternum convex, deeply sulcate behind; second ventral segment of abdomen impressed in the middle at apex, third minutely tuberculate at apex, these tubercles more or less approximate. Length 2·50-2·60 mm.

Hab. Natal (Frere), Zanzibar mainland (Lindi).

I had at first considered this species as identical with *T. collaris*, Raffr., and the description of the male which I gave in the 'Revue d'Entomologie,' 1887, p. 28, refers to this species and not to *T. collaris* proper; I deem it advisable therefore to give a comparative description of the two species.

T. collaris.	*T. rugicollis.*
Head longer, slightly narrowed before the antennal tubercle, which is but little dilated, foveæ between the eyes smaller than in *rugicollis*.	Head shorter, much narrowed before the antennal tubercle, which is much more dilated; foveæ between the eyes larger and much more transverse, separated from each other by a carinule.
Canthus under the eye produced in a long and sharp spine.	Canthus under the eye short and obtuse.
Last joint of maxillary palpi narrowly produced outwardly at apex, long, and very acuminate.	Last joint of maxillary palpi roundly dilated outwardly, apical part short, acuminate.
Male: Last joint of antennæ with only one large depression at base, ending at tip in a single, more or less compressed tooth.	Male: Last joint of antennæ with two transverse and compressed teeth, or small carinæ, the interval between them filled by a transverse depression.
Second and third ventral segments very slightly impressed.	Second ventral segment with a strong impression, third one with two small tubercles at base.
Length 2·40–2·60 mm.	Length 2·50–2·60 mm.
Hab. Island of Zanzibar, and on the mainland, Bagamoyo and Mikindani.	*Hab.* Natal (Frere), Zanzibar mainland (Lindi).

Gen. PSELAPHOCERUS, Raffr.,
Rev. Entom., vi., 1887, p. 28.

Thick, convex, attenuate in front; head with a broad, robust antennal tubercle; antennæ strongly and irregularly clavate, first joint subelongate, cylindrical, smaller than the following ones; palpi strong, first joint inconspicuous, second rather elongate, slender at base, much dilated in a triangular form, third triangular, more or less dilated outwardly either obtusely or in an apophyse, fourth more or less transversely triangular, and acuminate inwardly at tip; prothorax longer than broad, narrowed in front and foveate laterally; elytra large, convex, attenuate at base, ampliate behind, discoidal stria short; abdomen large, marginate, first dorsal segment hardly longer than the others; metasternum long, slightly raised; posterior coxæ hardly approximate; legs somewhat elongate, tarsi elongate, second and third joints subequal, two equal claws.

This genus is purely South African, and includes some of the large species of *Pselaphidæ*. It can be divided into two groups, containing each two species, and in each one of these two groups the maxillary palpi are exactly similar.

FIRST GROUP.

Third and fourth joints of the maxillary palpi obtusely produced outwardly; seventh joint of antennæ very large, ninth and tenth much smaller and equal .. .. .. *peringueyi, diversus*.

SECOND GROUP.

Third and fourth joints of the maxillary palpi narrowly and sharply produced outwardly in the shape of an appendage, seventh joint of the antennæ much smaller than the others, while the ninth is larger than the seventh and tenth.. .. .. *heterocerus, antennatus*

In general appearance these insects are very closely allied to each other, and the maxillary palpi being identical in the same group, it is rather difficult to distinguish the female of different species, while the difference in the shape of the antennæ makes the identification of the male comparatively easy.

PSELAPHOCERUS PERINGUEYI, Raffr.,
Plate XVII., fig. 10.
Rev. Entom., 1887, p. 29, pl. xvii., figs. 10 & 11.

Stout, attenuate in the anterior part, rufous, elytra lighter in the disk, shining, smooth, hirsute, the hairs brown and long, palpi testaceous; head subquadrato-elongate, plane, with two minute foveæ placed behind the eyes, middle of frontal part slightly impressed; antennæ robust, thick, first joint elongate, cylindrical, second nearly quadrate, third to fifth longer and slightly increasing in length, sixth quadrate, slightly transverse, the others different in each sex; prothorax larger than the head, irregularly ovate, somewhat abruptly attenuate in front, slightly dilated before the median part, and having laterally, a little past the median part, a transverse fovea filled with whitish glandulose hairs; elytra very much attenuate at base, shoulders well defined and oblique, base with two foveæ, filled at bottom with whitish glandulose squamæ, dorsal sulcus slightly oblique, short; abdomen of nearly the same size as the elytra; metasternum little impressed; trochanters and anterior femora strongly but obtusely tuberculate.

Male: Narrower behind; seventh joint of antennæ large, produced outwardly and rounded, eighth minute, lenticular, ninth to tenth large, lenticular, eleventh truncate at base, short, obtuse at tip, depressed, transversely excavate inwardly, and incised laterally also inwardly.

Female: Broader behind; seventh joint of antennæ a little larger than the others, subquadrate, eighth minute, transverse, ninth to tenth large, transverse, eleventh briefly ovate and obtuse at tip. Length 2·70–3 mm.

Hab. Cape Colony; somewhat common in Cape Town and neighbourhood—Newlands, Constantia.

Pselaphocerus diversus,
Plate XVII., fig. 11.

Nearly similar to the preceding species; the colour is more uniform and generally darker.

Male: Seventh joint of antennæ large, irregularly trapezoidal, hardly produced outwardly and rounded for a short space, eighth minute, transverse, ninth and tenth large, transverse, eleventh ovate, truncate at base, obtusely acuminate at tip, obliquely and broadly excavate inwardly, and with an ante-apical and squamose, longitudinal irregular carinule.

Female: Seventh joint of antennæ hardly broader than the preceding one, but nearly twice as long, eighth a little smaller than the sixth, and more transverse, ninth to tenth large, slightly transverse, eleventh ovate, obtusely acuminate; size of the male.

In both sexes the eighth, ninth, and tenth joints of the antennæ are much less transverse than in the male of *P. peringueyi*, the seventh is somewhat quadrate, whilst in *peringueyi* it has a triangular shape, the last one is neither so short or so broad, and the sculpture of the under part is quite different; in the female the seventh joint is longer but not broader, whilst in *peringueyi* it is broader and nearly quadrate, the eighth, ninth, and tenth are much less transverse, and the last one regularly ovate.

This species is much rarer than the preceding one. I have found only a few specimens near Cape Town on the Kloof Road, on the Camp's Bay side.

Pselaphocerus heterocerus, Raffr.,
Plate XVII., fig. 12.
Rev. Entom., 1887, p. 30, pl. i, figs. 8 & 9.

Suboblong, rufous, shining, smooth, body covered with long brown hairs, palpi testaceous; head elongate, somewhat plane, slightly constricted in the anterior part towards the antennal tubercle, which is subdivided, and having behind the median part two minute foveæ and two oblique very nearly obliterated sulci; eyes set towards the middle; first joint of antennæ long, cylindrical, slightly sinuate, second elongato-quadrate, third to fifth elongate, subcylindrical and slightly increasing in length, sixth quadrate, the others different in each sex; prothorax oblong, rather abrupt and more attenuate in front, nearly straight laterally, outer sides with a large, transverse fovea filled with whitish glandulose hairs; elytra rather elongate, slightly attenuate at base, shoulders oblique and well pronounced, two foveæ filled with whitish glandular hairs at base, dorsal sulcus short; abdomen shorter than the elytra; metasternum rather convex, hardly impressed; anterior trochanters and femora tuberculate.

Male: Seventh joint of antennæ a little longer than the preceding one, obtusely but strongly produced outwardly at apex, eighth not narrower than the sixth, but lenticular, ninth large, subquadrate, slightly transverse and rounded at angles, tenth much smaller, tranverse, eleventh rather elongate, truncate at base, acuminate at apex, slightly compressed, slightly produced inwardly and incised before the base, depressed inwardly and deeply excavated in a sub-rotund way.

Female: First to fifth joints of antennæ a little shorter than in the male, sixth to seventh quadrate, eighth transverse, ninth to tenth larger than the others, transverse, eleventh ovate and obtusely acuminate. Length 2·90 mm.

This species is very rare.

Hab. Cape Colony (environs of Cape Town, Constantia), Stellenbosch.

PSELAPHOCERUS ANTENNATUS,
Plate XVII., fig. 13.

This species is very similar to the preceding one, and differs only in the following points: The head is not quite so long, and the foveæ are larger but nearly obliterated; the first to fifth joints of antennæ are a little shorter and thicker; the prothorax has a well-defined notch on each side before the median part.

Female: Sixth joint of antennæ tranverso-quadrate, seventh more than twice larger than the preceding one, transverso-quadrate and slightly produced externally at apex, eighth lenticular, ninth large, transverse, rounded externally, tenth a little narrower and slightly transverse, eleventh truncate at base, attenuate at tip and obtuse, depressed inwardly and with the side almost roundly excavated.

Female unknown.

Hab. Cape Colony (neighbourhood of Cape Town, Newlands). Very rare; one example.

Gen. MARELLUS, Motschulsk.,
Bullet. Mosc., 1851, iv., p. 483.

Body oblong, attenuate in front; head transverse, antennal tubercle long, narrow; eyes large; maxillary palpi elongate, first joint conspicuous, second elongate, clavate at tip, third shorter, more or less obconical, fourth elongate, fusiform or filiform; antennæ elongate, club large, triarticulate; prothorax cordate and provided with a transverse sulcus and lateral foveæ; abdomen marginate; legs elongate, slender, posterior coxæ distant; tarsi elongate, second to third joints subequal, claws double, equal.

This genus seems to be exclusively African, and is found in Algeria, Egypt, Abyssinia, Zanzibar, and Natal, but examples are always rare.

Marellus granosus.

Totally rufous, granular, covered with an extremely short pubescence; head convex, transverse, antennal tubercle narrow and moderately elongate, sulcate longitudinally; third joint of palpi obconical, fourth in the shape of a long spindle, acuminate; antennæ elongate, first joint elongate, cylindrical, second a little thicker than the following ones, thicker, elongate and cylindrical, third to seventh elongate, subcylindrical, a little decreasing in length, eighth quadrate, ninth to tenth oblong, slightly obconical, tenth a little thicker and shorter, eleventh oblong, obtusely acuminate; prothorax cordate, abruptly attenuate in front, rounded laterally before the median part, and sinuate from there, two large lateral foveæ, the median one of which is minute, transverse sulcus straight, obsolete; elytra elongate, slightly attenuate in front, slightly oblique laterally, shoulders oblique and well defined, base bifoveate, sutural stria entire, dorsal sulcus short; abdomen shorter than the elytra, broadly marginate, the two first segments equal; metasternum strongly and broadly excavated, minutely carinate laterally, intermediate coxæ tuberculate inwardly; legs slender, femora slightly thickened; tibiæ not quite straight. Male. Length 1·40–1·50 mm.

In this species the last joint of the maxillary palpi is fusiform, as in *M. biskrensis*, but more acuminate; the prothorax is longer, less broad and less rounded in the anterior part; the elytra have more raised and more oblique shoulders; the metasternum is much more excavated and carinate on each side. It differs much more from *M. filipalpus*, Raffr., from Kilwa (Zanzibar mainland), in which the last joint of the palpi is elongate, slender, and filiform, the antennal club longer and the frontal tubercle shorter and thicker; the elytra are also more convex. In shape it resembles much *M. palpator*, Raffr., from Abyssinia, in which the elytra are also long and coarsely granulated, but in this last-named species the last joint of the palpi is cylindrical and slender, the prothorax shorter, and the antennal club longer.

Hab. Natal (Frere). One male example.

Gen. PSEUDOTYCHUS.

Body globose, attenuate in front; head subtriangular, provided with an antennal tubercle; eyes set behind; palpi moderate, first joint inconspicuous, second arcuate and slightly clavate, third small, fourth much larger, ovate, slightly securiform, acuminate; antennæ not quite approximate at base, club triarticulate; prothorax cordate, trifoveate and sulcate transversely; elytra convex, much attenuate at base and bifoveate, sutural stria entire, dorsal one wanting; abdomen convex, marginate, first dorsal segment larger than the others; metasternum convex, first ventral segment hidden by the metasternum, second large; legs of moderate size, posterior coxæ very far apart, intermediate trochanters a little elongate, but the insertion of the femora is, however, at the apex, the others short; first joint of tarsi minute, second produced inwardly, third much longer, claws double and equal.

This genus is very interesting; the intermediate trochanters are shorter than usual in this tribe, but they are nevertheless clavate, and the femur is inserted at the apex; the palpi are like those of *Bryaxis*, and the general facies is very much like that of the European genus *Tychus*, from which it is far removed by the elongate intermediate trochanters, the first ventral segment hidden by the metasternum and the two equal tarsal claws. It undoubtedly belongs to the *Tyrini* tribe, but it is an aberrant form.

Pseudotychus nigerrimus,
Plate XVII., fig. 5.

Black, shining, set with long but very sparse hairs; antennæ, palpi, and legs testaceous; head convex, subtriangular, antennal tubercle quadrate, transverse; antennæ robust, the first two joints a little larger than the others, third to seventh a little longer than broad, subcylindrical, eighth quadrate, ninth to tenth larger, quadrate, slightly transverse, eleventh ovate, truncate at base, obtusely acuminate at tip; prothorax much broader than the head, strongly cordate, more attenuate in front than behind, and having three foveæ of which the median is the smallest, and joined by an angular, transverse sulcus; elytra moderately short, more than twice as broad as the prothorax at apex, much attenuate at base, rounded laterally, shoulders obliterated, base bifoveate; abdomen larger than the elytra; femora slightly thickened, tibiæ straight, slightly clavate at tip; metasternum transverse, slightly convex, foveate close to the intermediate coxæ, broad at apex and truncate in a straight line; last ventral segment a little longer than the preceding one, and obsoletely impressed transversely. (Sex uncertain.) Length 1·40 mm.

Hab. Cape Colony (Cape Town neighbourhood, Newlands). Three examples.

Sub-Family CLAVIGERIDÆ.

The *Clavigeridæ*, in spite of their peculiar facies, which makes them easily distinguishable, even by beginners in entomology, from the true *Pselaphidæ*, do not in fact differ very much, and they appear to be a degraded form of the latter. However, the head, antennæ, mouth, abdomen, and sternum show some difference; the head is more or less conical with the lateral part of the epistoma always more or less dilated laterally; the antennæ are thick, short, and comprise never more than six joints, the first of which is concealed under the frontal part in a lateral fovea in which the antenna is inserted, the ultimate joint is always larger than the others, and, except in very few cases, truncate at tip; the mouth is very rudimentary and consists of long fascicles of soft hairs adapted for suction; the maxillary palpi are not visible, and have been found to consist of one joint only in such genera as have been dissected; the first dorsal segments of the abdomen coalesce, so that it has three dorsal segments on the upper part, while underneath it consists of the normal number; the intermediate coxæ are always apart, the metasternum being produced between them; this is a somewhat abnormal feature, because when the intermediate coxæ are apart, which is seldom the case in the *Pselaphidæ*, it is the mesosternum which is produced backwards between the coxæ; the posterior coxæ are always very broadly separated, and the trochanters of all the legs are very long and clavate, the femora being inserted on the apex of the trochanters and very remote from the coxæ; the base of the abdomen is always more or less, but generally very much, excavated, and each side is provided with large fascicles of hairs; these fascicles are always connected with tegumentary glands secreting a liquid of which, it is supposed, the ants, among which the *Clavigeridæ* are always found, are very fond. Some of these insects (*Claviger, Adranes*) are entirely eyeless.

Clavigeridæ are rare, but they occur in every part of the world, and are found in ants' nests; at times, however, they leave the formicaria at sunset, and are found climbing on plants in meadows.

In Madagascar some of the most curious species live in the nests built in trees by an ant (*Cremastogaster* spec.). I am sure that a methodical and careful search in ants' nests in South Africa will lead to the discovery of many more species than those known to occur.

Synopsis of Genera.

A 2. Antennæ four-jointed.
  B 2. Head broader and rounded at tip, hind margin of the abdominal excavation tridentate; second and third dorsal segments comparatively large .. .. *Fustigeropsis*.
  B 1. Head more or less acuminate at tip; abdominal excavation simply tuberculate or fasciculate on each side; second and third dorsal segments small.
    C 2. Abdominal excavation not tuberculate, and having on each side a fascicle of hairs connected with a larger fascicle at the external angle of the elytron, the apical margin of which is only slightly oblique .. .. .. *Novoclaviger*.
    C 1. Abdominal excavation having on each side a very strong, long, compressed tubercle fitting in a fasciculated notch on the external angle of the elytra .. .. *Fustigerodes*.
A 1. Antennæ five-jointed.
  Head short, acuminate in front; abdominal excavation fasciculate and strongly carinate on each side .. .. *Commatocerodes*.

Gen. FUSTIGEROPSIS, Raffr.,
Rev. Entom., 1890, pp. 164 and 167.
Commatocerus, Raffr., olim.

Antennæ quadri-articulate, elongate, last joint long, clavate at tip; head broader in front and rounded, anterior part of cheeks dilated in a triangular form; posterior margin of antennæ slightly marginate in the middle and fasciculate; abdomen long, first dorsal segment deeply impressed at base and tridentate, second and third conspicuous; legs somewhat elongate.

This genus contains only one species.

Fustigeropsis Peringueyi, Raffr.,
Plate XVII., fig. 8.

Commatocerus Peringueyi, Raffr., Rev. Entom., 1887, p. 19, pl. i., fig. 7.

Elongate, chestnut brown or rufo-testaceous, sparsely pilose; head elongate, very finely shagreened, sinuate in front of the eyes, ampliate and rounded in front, deeply foveate on each side; antennæ once and a half longer than the head, first joint hardly conspicuous, second subtransverso-quadrate, third obconical and a little longer than broad, fourth elongate, cylindrical, clavate at tip and truncate; prothorax cordate, very finely shagreened and having a few dispersed punctures; median part of base foveolate; elytra with remote, granulose hair-bearing punctures, attenuate at base, well developed at the shoulders, having a hardly conspicuous sutural stria and two short folds connected with the base and diverging, posterior margin sinuate, slightly incised and fasciculate in the median part; abdomen longer than the elytra, attenuate behind, slightly narrower at base, first segment large, marginate, deeply excavate transversely at base, the excavation with a strong trilobate margin, the lateral lobes are slightly fasciculate, and the second and third moderately large; legs elongate, hardly thickened, simple. Length 2·20 mm.

The antennæ are more elongate, and the second and third dorsal segments much larger than in most of the *Clavigeridæ*.

I have seen two examples only (female), which are not in very good condition.

Hab. Transvaal (Potchefstroom).

Gen. FUSTIGERODES, Reitt.,
Deutsch. Ent. Zeit., 1884, p. 168.

Antennæ quadri-articulate, less elongate than in *Fustigeropsis*, last joint more or less cylindrical or slightly conical; head short, attenuate in front, anterior part of cheeks dilated but nearly straight, posterior margin of elytra more or less oblique and with a slightly nodose fascicule; first dorsal segment of abdomen deeply impressed transversely at base, and having on each side a large process, depressed at tip and hardly fasciculate, second and third hardly conspicuous on the upper side.

This genus has been established by Mr. Reitter for an undescribed insect, the habitat of which was not known, and which proved afterwards to be identical with *Commatocerus capensis*, Pér.

The genus *Commatocerus* has caused a good deal of confusion, which I have tried to remove in a critical discussion in the 'Revue d'Entomologie,' 1890, pp. 166 and 167.

Fustigerodes capensis, Pér.

Commatocerus capensis, Pér., Trans. S. Afric. Phil. Soc., 1888, p. 84; Raffr., Rev. Entom., 1890, pl. iii., figs. 19, 19[1].

Elongate, rufous or chestnut brown, sparsely setose; head thick, attenuate in front, very rugosely punctate, foveolate on each side at the back of the eyes; antennæ longer than the head, first joint inconspicuous, second minute, third transverse and a little narrower than the following one, fourth elongate, subcylindrical, slightly attenuate at base and equally slightly at apex, truncate; prothorax rounded, very rugoso-punctate, longitudinal sulcus strong and ending in a pre-basal large fovea, lateral foveæ large; elytra convex, suture depressed, sutural stria entire, broadly bi-impressed at base, discoidal stria delicate, arcuate and a little attenuate behind the middle, the other fold is short and subhumeral, the posterior margin is a little oblique, placed close to the external angle, subnodose and fasciculate; abdomen large, rather convex, somewhat rounded at apex, first dorsal segment excavated transversely at base, and provided on each side with a large, elongate process, depressed at tip legs strong, intermediate femora with a large, triangular tooth at base, intermediate tibiæ with an inward sharp tubercle placed at a long distance from the apex.

In this species the last joint of antennæ is cylindrical, attenuate at base, and slightly thicker about the median part, more especially if seen sideways.

Hab.   Cape Colony (Grahamstown).

FUSTIGERODES MAJUSCULUS, Pér.,
Plate XVII., fig. 7.

Commatocerus majusculus, Pér., Trans. S. Afric. Phil. Soc., 1888, p. 84.

Elongate, chestnut red, a little setose; head short, linear, thick, very rugose, punctured, the punctures subocellate, triangular and attenuate in front, deeply foveate behind the eyes; antennæ short, first joint inconspicuous, second subtransverse, third slightly obconical, fourth attenuate at base, straight, sensibly increased towards the apex, rugose and thickly setose; prothorax slightly transverse, more attenuate in front than behind, very rugose and punctured, punctures subocellate, median longitudinal sulcus strong and ending in a large basal fovea, lateral foveæ of moderately size; elytra of a lighter colour that the rest of the body, moderately elongate, convex, suture depressed, sutural stria entire, base strongly bi-impressed, discoidal stria, or rather fine fold, slightly arcuate and attenuate in the middle, posterior margin oblique, nodoso-fasciculate in the middle at a distance from the external angle; abdomen large, very convex, rounded a little laterally and a little broader than the elytra, deeply impressed at base and having on each side a ciliate, large, elongate process depressed at tip, rugose and bicarinate; legs robust.

Male: Intermediate femora excavate inwardly at base and armed with an inner very large, triangular, compressed tooth; intermediate tibiæ minutely dentate inwardly before the apex.

Length 2·40–2·60 mm.

This splendid insect differs from *F. capensis* by the larger size and the shorter antennæ, the last joint of which is slightly conical.

Hab. Cape Colony (neighbourhood of Cape Town, Newlands). Found in the nest of *Acantholepis capensis*.

Gen. NOVOCLAVIGER, Wasmann,
Kritisch. Verzeich. Myrmec. Termitoph. Arthrop., p. 214.

Antennæ quadri-articulate, first joint inconspicuous, second to third minute, fourth rather elongate, slightly clavate, truncate at tip; head slightly attenuate in the anterior part, cheeks dilated in a nearly straight line in front; elytra with the posterior margin entire and with the very external angles strongly fasciculate; first dorsal segment with a large, transverse, simple fovea, cicatricose and fasciculate on each side, second and third hardly conspicuous on the upper side.

Closely allied to *Fustigeroides*, Reitt.; the antennæ are similar to those of *F. majusculus*, but the apical part of the elytra is simple and slightly oblique, whilst the external angle itself is fasciculate; the lateral processes of the abdomen are wanting, and replaced by a fascicle of hairs.

Novoclaviger wroughtoni, Wasm.,
Plate XVII., fig. 6.
Loc. cit., p. 215.

Elongate, rufo-testaceous, body covered with moderately dense, flavous, setose hairs; head and prothorax closely punctate, the punctures very rough; head subcylindrical, slightly attenuate in front and foveate laterally beyond the eyes; antennæ a little longer than the head, second joint minute, third twice as long as the preceding one and longer than broad, fourth elongate, slightly clavate, truncate at apex; prothorax nearly globose, more attenuate in front than behind, canaliculate longitudinally, lateral foveæ and also the median one moderately large; elytra longer than broad, very slightly rounded laterally, base with two folds, apical margin deflexed and slightly oblique, external angles raised and with conspicuous golden-yellow fascicles of hairs; abdomen a little longer than the elytra, rounded laterally, disk subglobose, the whole base impressed transversely, cicatricose on each side close to the margin and fasciculate; legs of moderate size.

Male: Abdomen impressed longitudinally underneath, femora of the intermediate legs with a triangular tooth of moderate size underneath, anterior tibiæ with an extremely small tooth set before the apex. Length 1·90 mm.

Wasmann states, *loc. cit.*, that this insect has been discovered in Delagoa Bay by Mr. C. Wroughton; the habitat, however, does not seem to be a correct one. Mr. Wroughton, of the Indian Forest Department, while on a visit to the Cape and Natal, collected a good number of ants, and sent to the South African Museum some duplicates of his captures, among which was one example of the *Novoclaviger*, above described, and said by him to have been found in Natal, examples of which had been forwarded to Wasmann. There can be no doubt about the identity of the insect sent by Mr. Wroughton to the South African Museum and Wasmann's type, as I have received from the latter one of his types which I have compared with the example in the South African Museum, and they are absolutely identical. The habitat of the insect is therefore Natal, in all probability, and not Delagoa Bay.

Gen. COMMATOCERODES, Pér.,

Trans. S. Afric. Phil. Soc., 1888, p. 85; Raffray, Rev. Entom., 1890, pp. 164–168.

Oblong, rather thick, antennæ elongate and five-jointed, first joint inconspicuous, second transverse, third to fourth subquadrate, fifth much elongate, clavate; head short, triangular in the anterior part, and acuminate, cheeks strongly dilated in front; margin of elytra simple; first dorsal segment of abdomen very much impressed transversely at base, strongly plicate and fasciculate on each side, second and third conspicuous on the upper part.

This genus is very different from the preceding one owing to the antennæ, which are really five-jointed, and much longer, the short and broad head, and the strongly plicate abdominal excavation, fasciculate on each side

Commatocerodes Raffrayi, Pér.,

Loc. cit., p. 86, pl. 1, fig. 3; Raffray, Rev. Entom., 1890, pl. iii., fig. 26, 26[1].

Oblong, short, and thick, chestnut brown, very briefly setulose; head and prothorax closely and roughly punctured, the punctures subocellate; head short, broad, thick, triangular and acuminate at tip, obsoletely foveate laterally behind the eyes, which are set backwards; antennæ hardly shorter than the head and prothorax together, fourth joint rugose, setose, clavate at tip, truncate and slightly sinuate outwardly; prothorax hardly longer than the head, transverse; elytra very minutely shagreened and punctulate, a little broader than long, apical margin straight and simple, external angles long, setose, sutural stria entire, triplicate at base; abdomen longer than the elytra and a little broader at base, first dorsal segment very much excavated transversely at base, the excavation carinate on each side and fasciculate, lateral margin broader at base and slightly incised externally, narrowed at tip; second segment conspicuous, transverse, and having a minute tubercle on each side of the base, third segment longer than the others, apex elongate and curving upwards; legs of moderate size, simple, somewhat elongate and setose. Female. Length 2·50 mm.

This species is a very singular one. The only example known is a female, and the darker patch of the apical part of the abdomen, mentioned by Mr. Péringuey in his description, is caused by the dried-up ovaries which are seen through the transparency of the teguments; a case occurring often in all light-coloured *Pselaphidæ*. I presume that the recurved protuberance of the last dorsal abdominal segment is also a characteristic of the female.

Hab. Transvaal (Potchefstroom).

POSTSCRIPT.

[The species described below belongs to the genus *Dalmina*, Raffr. See page 78.]

DALMINA ELIZABETHANA.

Oblong, moderately thick, bright red with the last joint of antennæ lighter in colour, and covered with a short, depressed, pallid pubescence; head longer than broad, slightly attenuate in front, and having between the eyes two large foveæ and two sulci slightly arcuate and joined in the anterior part; antennæ with long setæ little elongate, first joint short, thick, second a little shorter, quadrate, third narrower by nearly half and a little longer than broad, fourth scarcely longer but nearly twice as broad, irregularly transverse and slightly produced inwardly, fifth large, irregularly transverse, with the internal basal angle produced underneath and mucronate, sixth much smaller than the fourth and transverse, seventh equal in width to the preceding one, but longer and less transverse, eighth similar to the sixth, sixth, seventh, and eighth joints slightly produced inwardly, ninth and tenth hardly smaller than the preceding one and transverse, eleventh hardly broader, truncate at base, acuminate at apex and slightly turbinate; prothorax broader than the head, cordate, sinuate laterally past the middle, lateral foveæ large, median transverse sulcus much angulate, median fovea almost wanting; elytra much broader than the prothorax, longer than broad, slightly rounded laterally, shoulders oblique, little prominent, base with two foveæ, dorsal sulcus disappearing before reaching the median part; metasternum minutely foveate in the posterior part; legs of moderate size, posterior trochanters slightly produced and aculeate at apex, posterior tibiæ a little thickened and with an apical spur; last ventral segment neither broadly nor deeply impressed. Male. Length 1·80 mm.

Closely allied to *D. globulicornis*; the fourth joint of the antennæ is also thicker than the fifth, although not quite as much thickened, but these two joints do not unite to form a knob as in *D. globulicornis*, and the shape of the fifth joint is more rounded; it differs from *D. irregularis* in the form of the fourth and fifth joints, which are transverse instead of being elongato-quadrate; it cannot be mistaken for the male of *D. concolor* because the head is longer than broad, whereas in the latter it is transverse, and the joints of the antennæ are longer than broad; it cannot be compared with *D. gratitudinis*, the fifth joint of which is much more dilated, while the fourth is hardly so. In short, in both *D. globulicornis* and *D. elizabethana* two joints, the fourth and fifth, are dilated, but in the former these two joints form a common, slightly rounded node by the superposition of the anterior and posterior margins respectively, whereas in the latter these two joints are very free and distant from one another; the colour is also much darker in *D. globulicornis*. Female unknown.

Hab. Cape Colony (Port Elizabeth).

INDEX TO FAMILY PSELAPHIDÆ.

A

	PAGE
abdominalis (Raffrayia)	74
achillis (Reichenbachia)	96
afra (Reichenbachia)	94
africanus (Faronidius)	47
Anoplectus	59
antennatus (Pselaphocerus)	112
armata (Raffrayia)	69
Arthmius	85
Asymoplectus	55
aterrimus (Asymoplectus)	58
atratus (Asymoplectus)	59
australis (Ctenistes)	103

B

Batoxyla	83
Batrisini	82
Batrisodes	86
Batrisus	84
brevipennis (Laphidioderus)	102
bicolor (Raffrayia)	71
Bryaxinini	87
Bryaxis	89

C

calcarata (Raffrayia)	68
capensis (Fustigerodes)	117
capensis (Laphidioderus)	101
caviceps (Raffrayia)	64
caviventris (Asymoplectus)	57
circumflexa (Rybaxis)	88
Clavigeridæ	115
Commatocerodes	120
concolor (Dalmina)	80
costatus (Odontalgus)	106
cruciata (Raffrayia)	70
Ctenistes	103
Ctenistini	100

D

Dalmina	78
decipiens (Reichenbachia)	92
deplanata (Raffrayia)	65
discicollis (Asymoplectus)	55
discoidalis (Euplectus)	53
discreta (Reichenbachia)	93

	PAGE
diversa (Reichenbachia)	94
diversus (Pselaphocerus)	111
dregei (Trabisus)	84

E

elegans (Dalmina)	82
elizabethana (Dalmina)	121
Euplectini	49
Euplectus	53

F

Faronidius	47
Faronini	47
Fustigerodes	117
Fustigeropsis	116

G

globulicornis (Dalmina)	79
Goniacerini	97
granosus (Marellus)	113
gratitudinis (Dalmina)	79

H

hæmatica (Bryaxis)	90
heterocera (Xenogyna)	61
heterocerus (Pselaphocerus)	111

I

imitator (Ctenistes)	104
incerta (Raffrayia)	65
irregularis (Asymoplectus)	56
—— (Dalmina)	81

L

laticeps (Zethopsus)	50
laticollis (Raffrayia)	73
Laphidioderus	100
longiceps (Pselaphus)	99
longula (Raffrayia)	78
luctuosus (Asymoplectus)	57

M

majorina (Raffrayia)	70
majusculus (Fustigerodes)	118
Marellus	113
mashuna (Syrbatus)	86
microcephala (Raffrayia)	76

Index.

N

	PAGE
nasuta (Raffrayia)	67
natalensis (Batrisodes)	87
——— (Raffrayia)	75
niger (Anoplectus)	60
nigerrimus (Pseudotychus)	114
nodosa (Raffrayia)	75
Novoclaviger	119

O

obscura (Raffrayia)	77
Odontalgus	104
Ogmocerus	97

P

palustris (Trimyodites)	52
peringueyi (Fustigeropsis)	116
——— (Pselaphocerus)	110
——— (Reichenbachia)	95
picticornis (Reichenbachia)	91
pilosella (Raffrayia)	74
Prodalma	51
Pselaphidæ	43, 46, 47
Pselaphini	98
Pselaphischnus	99
Pselaphocerus	109
Pselaphus	98
Pseudotychus	114
punctata (Batoxyla)	83

Q

quadriceps (Euplectus)	54

R

	PAGE
raffrayi (Commatocerodes)	120
Raffrayia	62
Reichenbachia	90
rugicollis (Tmesiphorus)	108
rugosula (Raffrayia)	72
rugosus (Ogmocerus)	98
Rybaxis	88

S

squamosus (Pselaphischnus)	99
subpubescens (Reichenbachia)	92
sulcicollis (Zethopsus)	50
sulcicornis (Reichenbachia)	90
Syrbatus	85

T

Tmesiphorus	107
Trabisus	84
Trimiodytes	52
Tyrini	107

V

variabilis (Raffrayia)	66
vespertinus (Odontalgus)	105

W

wroughtoni (Novoclaviger)	119

X

Xenogyna	60

Z

Zethopsus	49

BIBLIOGRAPHY TO FAMILY PSELAPHIDÆ.

Aube, C.
  Psclaphorum Monographia cum Synonymia extricata. *Magas. d. Zool.*, 1833. Nos. 78–94, *pp.* 71, 17 pl.
  Révision de la famille des Psélaphiens. *Annal. Soc. Entom. d. France*, 1844, *pp.* 73–156.

* Herbst, J. F. W.
  Natursystem aller bekannten in- und auslandischen Insecten, 1785–1806, 8vo, 2 vols. Käfer, vols. i.–x.

Leach, W. E.
  On the Stirpes and Genera composing the family Pselaphidæ. *Vigor's Zoolog. Journ.*, 1825, vol. ii., *pp.* 445–453.
  The Zoological Miscellany. London, 1814–1817, 3 vols. 8vo.

Le Conte, J. L.
  On the Pselaphidæ of the United States. *Boston Journ. Nat. Hist.*, 1850, vol. vi., *pp.* 64–110.

Motschulsky, V. de
  Enumération des nouvelles espèces de Coléoptères rapportées par Mr. de M. de son dernier voyage. *Bull. d. Mosc.*, 1851, iv., *pp.* 479–511.

* Péringuey, L.
  Second Contribution to the South African Coleopterous Fauna. *Trans. S. Afric. Phil. Soc.*, vol. iv., 1888, *pp.* 67–194.

Raffray, A.
  Description d'espèces nouvelles de la famille des Psélaphides. *Revue et Magas. d. Zoologie*, 1877, *pp.* 279, 298, 1 pl.
  Psélaphides nouveaux ou peu connus. *Revue d'Entom.*, i., 1882, *pp.* 1–16, 25–40, 49–64, 73–85, pls. i. and ii.
  Psélaphides nouveaux ou peu connus. *Revue d'Entom.*, vol. vi., 1887, *pp.* 18–56, 61–62, 2 pl.
  Etudes sur les Psélaphides. *Revue d'Entom.*, vol. ix., 1890, *pp.* 1–28, 81–220, 264–265, 3 pl.
  Voyage de Mr. E. Simon dans l'Afrique Australe: Psélaphides. *Annales Soc. Entom. de France*, 1895, *pp.* 389–390.

Reichenbach, H. S. L.
  Monographia Pselaphorum. Lipsiæ, 1816, 8vo, *pp.* 79, 2 pl.

Reitter, Ed.
  Beitrags zur Kenntnifs der Pselaphiden und Scydmæniden von West Africa. *Deutsch. Entom. Zeit.*, xxvi., 1882, *pp.* 177–195, 2 pl.

Ueber die bekannten Clavigeriden Gattung. *Deutsch. Entom. Zeit.*, vol. xxviii., 1884, *pp.* 167–168.

Versuch ein. systemat. Eintheil. der Clavigeriden und Psclaphiden. *Verhand. d. naturforsch. Vereines in Brünn*, vol. xx., *pp.* 177–211.

SAULCY, F. DE
    Species des Paussides, Clavigérides, Psélaphides et Scydménides de l'Europe et des pays circonvoisins. *Bullet. Soc. d'Hist. Natur. de Metz*, xiv., 1876, *pp.* 25–100.

SCHAUFUSS, C.
    Catalogus synonymicus Pselaphidarum adhuc descriptarum. *Tijdschr. Ent.* xxxi., 1888, *pp.* 1–104.

WASMANN, E.
    Kritisches Verzeichniss der Myrmecophilen und Termitophilen Arthropoden. Berlin, 1884, 8vo, 231 *pp.*

ADDENDA TO FAMILY PSELAPHIDÆ.

Tribe EUPLECTINI.
(p. 49.)

Gen. TRIMIODYTES.
(p. 52.)

Trimiodytes setifer.

Oblong, subparallel, somewhat convex, obscurely ferruginous; antennæ and legs rufous; palpi testaceous; it is clothed with a hardly discernible pubescence, but has, however, some long erect scattered setæ. Head trapezoid, attenuate in front, with the frontal part truncate and slightly produced in the middle, bifoveate anteriorly, transversely sulcate and having between the eyes two large foveæ joined by a deep furrow to the transverse frontal sulcus; eyes of moderate size; antennæ also of moderate size, and with the two basal joints larger than the others, third to ninth inclusive moniliform and slightly decreasing in length, but with the fifth a little larger, tenth a little larger the preceding ones, slightly transverse, eleventh ovate, truncate at base, acuminate and subturbinate at apex; prothorax nearly smaller than the head, very cordiform, with the outer sides rounded before the median part and sinuate after the lateral fovea, which is large; the transverse sulcus is sinuate, and the median fovea smaller than the others; elytra elongate, subparallel, shoulders defined and oblique, base broadly bifoveate, sutural stria abbreviated before the median part; abdominal segments equal; legs of moderate size; anterior femora a little thickened; metasternum convex, simple; second ventral segment larger than the others, third to fifth inclusive decreasing, sixth large, transverse, convex. Female. Length 1·20 mm.

This second species of a genus, restricted hitherto to the Cape Peninsula, and which seems to take the place of the *Trimium* of Europe and the *Actium* of America, differs from *T. palustris*, Raffr., by its stouter and more parallel facies; the head is proportionally less large and is shorter, the frontal part is provided with a small obtuse projection, and the elytra have well-defined shoulders instead of being obliterated as in *T. palustris*.

Hab. Stellenbosch (Cape Colony). A single example captured in February on the banks of the river.

Tribe BATRISINI.
(p. 82.)

Gen. BATRISUS.
(p. 84.)

Subgen. APOBATRISUS, Raffr.,
Soc. Ent. d. France, 1896, p. 235.

This subgenus is distinguished from *Trabisus*, Raffr., and *Probatrisus*, Raffr., by the sensibly narrower smaller head, a little compressed laterally, and having underneath on each side a deep carina edging laterally a depression intended for the reception of the palpi, which are flagellate as in the two above-named subgenera, which character, coupled with the very large abdominal segment having no lateral carina, differentiates them from the true *Batrisus*, Aubé. The value of the prothoracic sulci as a character is not important because they vary; they are very well defined in *Batrisus* and *Trabisus*, which have three, and in *Probatrisus*, which has one only, and disappear almost entirely in *Apobatrisus*. In one species of the latter there is none; in another it is replaced by a sulciform fovea.

This subgenus was established for a species from Gaboon; but another species has lately been discovered in the Cape Colony.

Apobatrisus rufus.

Oblong, rufo-testaceous, entirely covered with rough punctures, and clothed with a brief but dense fulvous pubescence. Head narrower than the prothorax, attenuated in front, with the frontal part sulcate transversely, retuse behind, and having between the eyes two foveæ more distant from one another than from the eyes; antennæ short, robust, joints third to eighth inclusive decreasing in length, eighth nearly transverse, ninth and tenth larger than the others and transverse, eleventh ovate, acuminate at tip; prothorax nearly triangular and broadest near the anterior angles, slightly transverse, lateral foveæ placed at a great distance from the outer side, transverse sulcus hardly well defined, median fovea smaller than the others, disk with a sulciform fovea, base bifoveate; elytra broader than the prothorax, and longer than broad, outer sides slightly rounded, shoulders raised, dorsal sulcus deep but very short; abdomen slightly narrowed at base, rounded laterally, first segment very large, hardly carinulate laterally at the base, but trifoveate, with the median fovea transverse, more than twice as broad as the lateral ones, and limited on each side by a short carinule; legs short, moderately thick, all the tibiæ slightly thickened, the posterior ones incurved; last ventral segment large, simple; metasternum plane, obsoletely sulcate. Female. Length 2·20 mm.

This species is very distinct from *A. gabonicus*, Raffr. The colour is much lighter, the prothorax is broadest at the very apex, the transverse sulcus is almost lost, and the discoidal fovea sulciform. One example only.

Hab. Cape Colony (? Beaufort West).

Tribe BRYAXININI.
(p. 87.)

Gen. REICHENBACHIA.
(p. 90.)

Reichenbachia rivularis.

Dark castaneous; antennæ and legs lighter; palpi pale testaceous; body clothed with a greyish pubescence. Head of moderate size, a little longer than broad, attenuate in front, and having three equal foveæ; antennæ somewhat shorter and elongate, joints third to fifth inclusive elongate, sixth also elongate, but a little shorter, seventh longer than broad, eighth slightly transverse, ninth a little longer, quadrate, tenth wider, slightly transverse, eleventh suboblong, acuminate at tip, and truncate at base; prothorax larger than the head, much attenuate in the anterior part, and moderately so in the posterior, basal median fovea slightly oblong, base punctate; elytra large, attenuate at base, shoulders subnodose, base trifoveate, dorsal stria slightly arcuate inwardly and abbreviated before the apex; the carinules of the first abdominal segment are a little divergent, include the fourth part of the disk, and reaching to about one-third of the length; metasternum obtusely tuberculate at base close to the intermediate coxæ, and broadly depressed; last ventral segment paler, strongly foveate, the fovea subquadrate; trochanters simple, intermediate tibiæ very briefly spurred at apex, posterior ones slightly incurved and thickened towards the apex. Male. Length 1·80 mm.

This new species resembles the female of *R. diversa*, Raffr., and still more the male of *R. Peringueyi*, Raffr. In both these species the third to the fifth joints inclusive are only twice as long as broad, the sixth is hardly longer than broad, the seventh is quadrate, and the eighth transverse. In *R. rivularis* the joints three to five are three times as long as broad, the sixth twice as long as broad, and the seventh a little longer than broad; the abdominal striæ are as in *R. Peringueyi*, and not so distant as in *R. diversa*; the prothorax is less narrowed behind than in both these species. In comparison with the male of *R. Peringueyi*, it is of larger size, the metasternum is noticeable on account of the basal blunt tubercle, the fovea of the last ventral segment is larger, and the apical spur of the intermediate tibiæ is not so robust. It cannot be compared to the male of *R. diversa*, the last antennal joint of which is provided with an inward tooth, entirely absent in *R. rivularis*.

Hab. Cape Colony (Stellenbosch). One example, captured at the foot of grass growing on the banks of the river.

[The above additions to the family *Pselaphidæ* were too late to be included in the Index, pp. 123, 124.]

EXPLANATION OF PLATES.

Plate XVI.

1. Raffrayia caviceps, Raffr., 1¹ ♂ underside of the head.
2. ,, armata, Raffr., 2¹ ,, ,,
3. ,, variabilis, Raffr., ♂, ♀ 3 ,, ,,
4. ,, incerta, Raffr.
5. ,, deplanata, Raffr.
6. ,, calcarata, Raffr., antennæ, 6¹ intermediate tibia of ♂.
7. ,, cruciata, Raffr.
8. ,, majorina, Raffr.
9. ,, natalensis, Raffr., ♂.
10. ,, nasuta, Raffr., ♂ 10¹ underside of the head.
11. ,, nodosa, Raffr.
12. ,, microcephala, Raffr.
13. ,, longula, Raffr., 13¹ ♂ underside of the head.
14. ,, bicolor, Raffr.
15. ,, pilosella, Raffr.
16. ,, laticollis, Raffr.
17. Xenogyna heterocera, Raffr., ♂.
18. Erimiodytes palustris, Raffr.
19. Prodalma capensis, Raffr., ♂.
20. Batrisodes natalensis, Raffr., ♂, upper side of the abdomen.
21. Asymoplectus discicollis, Raffr., ♂ last ventral segment.
22. ,, caviventris, Raffr., ♂ ,,
23. ,, irregularis, Raffr., ♂ ,,
24. ,, atratus, Raffr., ♂ ,,
25. ,, aterrimus, Raffr., ♂ ,,
26. ,, luctuosus, Raffr., ♂ ,,

Plate XVII.

1. Dalmina elegans, Raffr., ♂.
2. Pselaphischnus squamosus, Raffr.
3. Syrbatus Mashuna, Raffr., ♂.
4. Batoxyla punctata, Raffr.
5. Pseudotychus nigerrimus, Raffr.
6. Novoclaviger Wroughtoni, Wasmann, ♂.
7. Fustigerodes majusculus, Péringuey, ♂.
8. Fustigeropsis Péringueyi, Raffr., ♀.
9. Odontalgus costatus, Raffr.
10. Pselaphocerus Péringueyi, Raffr., ♂ antenna.
11. ,, diversus, Raffr., ♂ ,,
12. ,, heterocerus, Raffr., ♂ ,,
13. ,, antennatus, Raffr. ♂ ,,

Pl. XII

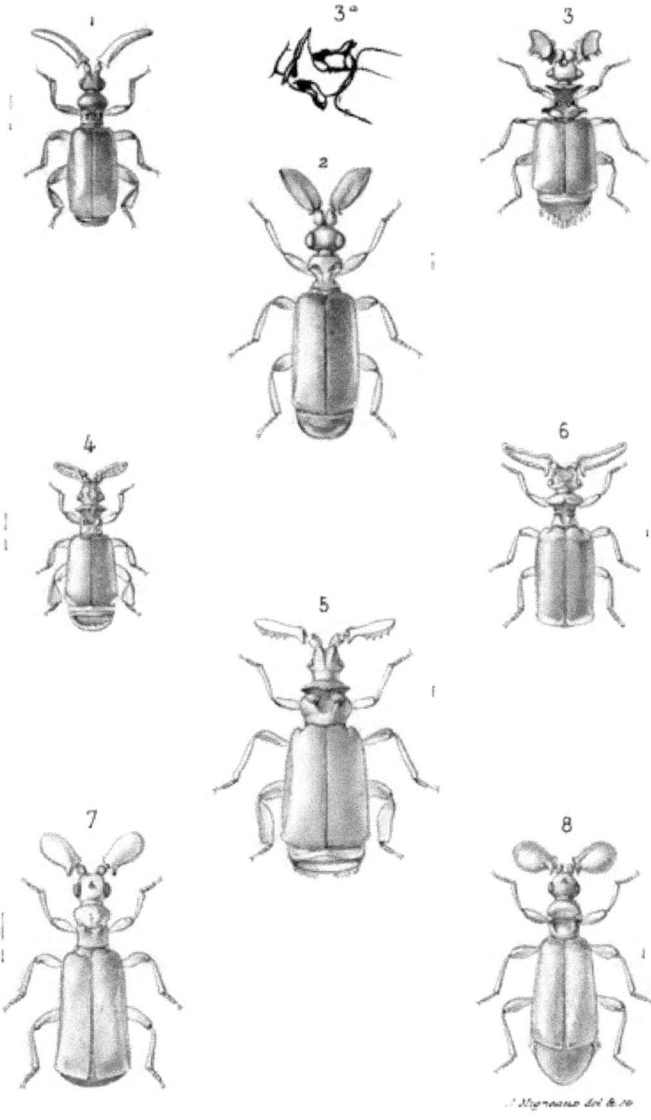

1	Paussus Raffrayi	5	Paussus Barberi
2	manicanus	6	concinnus
3	Marshalli	7	propinquus
4	viator	8	arduus

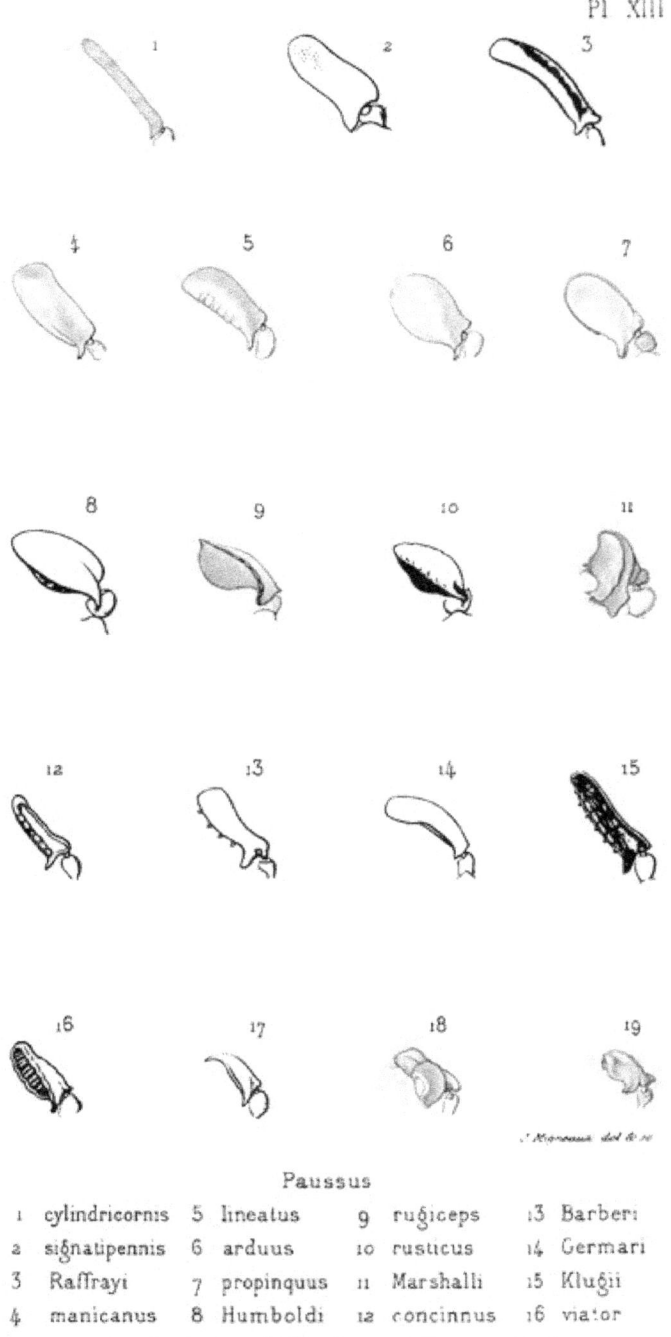

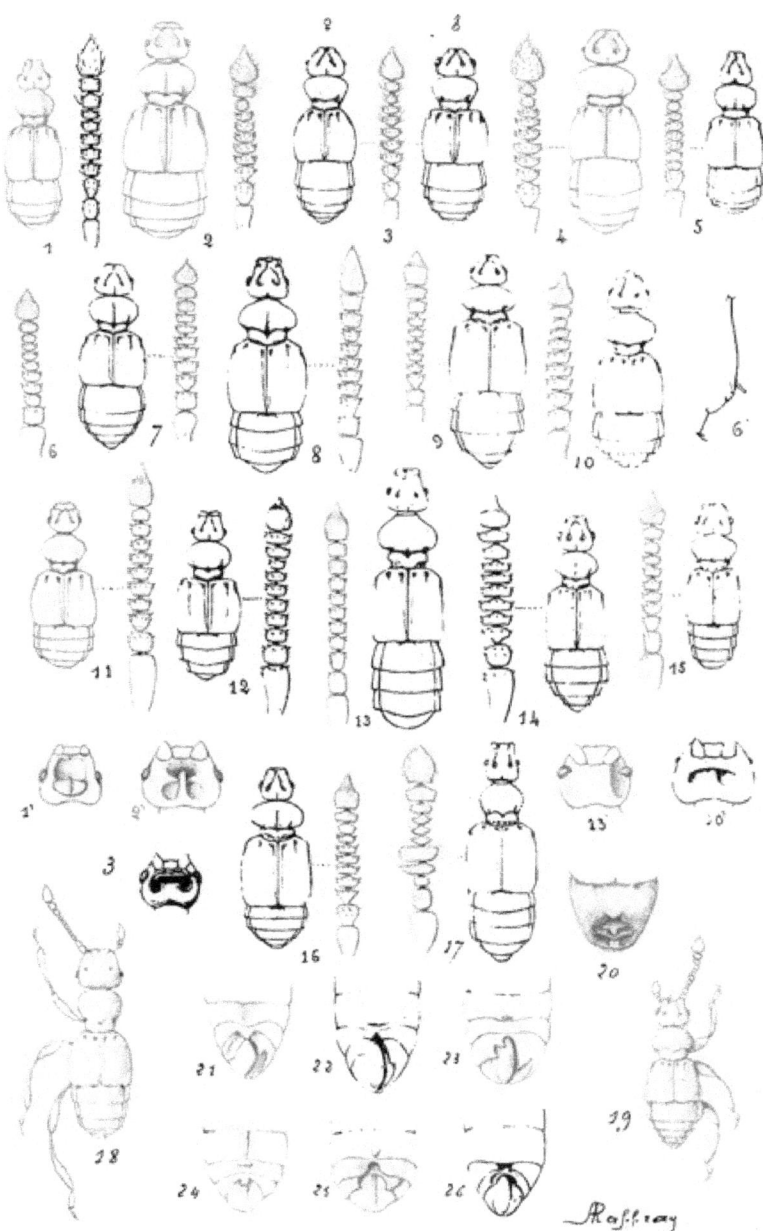

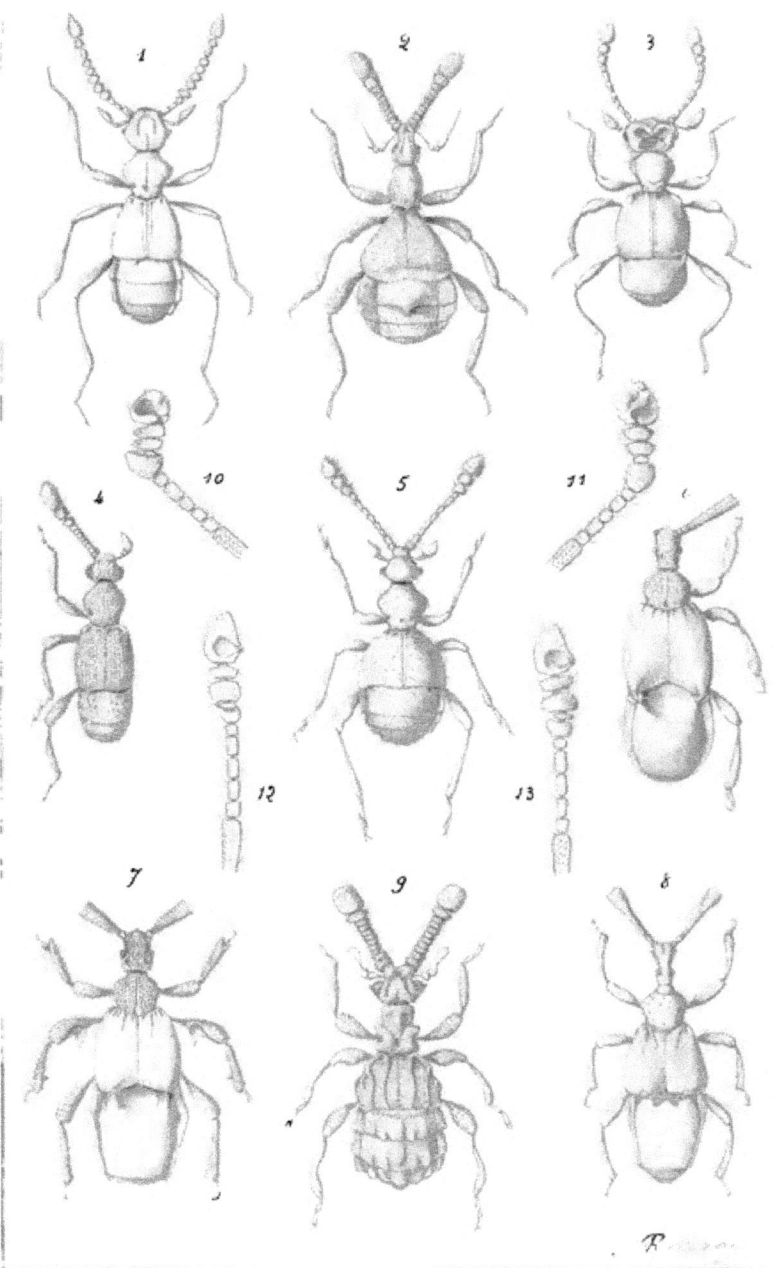

DESCRIPTIVE CATALOGUE OF THE COLEOPTERA OF SOUTH AFRICA.—Part IV.

By A. Raffray, M.E.S. France, &c.

Family PSELAPHIDÆ.

FIRST SUPPLEMENT.

With Plate 6 (XVIII.).

The publication of the Descriptive Catalogue of the South African *Pselaphidæ* in 1888 has encouraged the entomologists in that part of the world to collect these curious beetles, which, owing to their small size, escape the attention of the ordinary collector.

The Catalogue contained 80 species; I now add 26 new ones to this number. The accession is a considerable one. If one takes into consideration the fact that these minute Coleoptera have been collected in some isolated parts only of South Africa, *i.e.*, round Cape Town, Muizenberg, and Stellenbosch by Mr. Péringuey and myself, Port Elizabeth by Dr. Brauns, Uitenhage by Rev. J. A. O'Neil, Salisbury and Frere (Natal) by Mr. G. A. K. Marshall, one is justified in assuming that when they have been looked for methodically and systematically in this part of Africa they will number several hundred.

Many new genera and forms entirely unknown will certainly be discovered, but the material which enables me to publish this First Supplement goes far to corroborate the opinion I have already given, that the south-western part, and more especially the Cape peninsula, has a very distinct fauna, while the western part of the Colony, Natal and Zambesia, are more directly connected with the general African fauna.

This Supplement contains diagnoses of five genera not until now recorded from South Africa; two are entirely new (*Gabata, Bryaxonoma*), while three (*Pselaphoxys, Sognorus*, and *Centrophthalmus*) are known to occur in other parts of Africa. *Gabata* has been found at Port Elizabeth; *Bryaxonoma* is from Muizenberg, near Cape Town; *Pselaphoxys*, described at first from Abyssinia, has been met with at Uitenhage, and the species is even identical with the Abyssinian one, *Centrophthalmus*, found on the eastern and western coasts of Africa, and reaching northwards as far as Algeria, is represented by two species—one from Salisbury, Mashunaland, the other from Uitenhage, Cape Colony. *Sognorus*, which is spread on Europe, Asia, and America, and has also one representative on the African West Coast, is represented by one new species found in Uitenhage.

This last-named locality seems to be a connecting point between the South-Western fauna of the Colony and the African one, for there the genus *Trimiodytes*, which is exclusively South African, and the number of species of which seems to be on the increase, occurs together with *Sognorus*; but so far the genus *Raffrayia*, which has now 26 representatives, and *Pselaphocerus* which has 6, both of which are so characteristic of the *Pselaphid* fauna of that part of the world, have not as yet been met there. Port Elizabeth, on the other hand, has 4 species of *Raffrayia*, 1 *Pselaphocerus*, 2 *Trimiodytes*, and 1 *Fustigerodes*, and seems to have more affinity with the distinct fauna of the peninsula.

The division of the South African *Pselaphidæ* in two faunas, although so interesting, is not possible yet, and I do not know that it can ever be a very precise one; there will always be found species which for one reason or other have a very wide area of geographical distribution, and there will always be points where the two faunas will commingle, yet my opinion is that the study of these insects, taken as a whole, will confirm the division in two faunas—one restricted to the South-Western region, and peculiar to it, the other spread on the Northern and Eastern side, and having a close affinity to the general African fauna.

Tribe FARONINI.

Gen. FARONIDIUS, Casey,

Catal., p. 47.

FARONIDIUS MONILIS.

Moderately elongate, rufous or testaceous, antennæ and legs testaceous, covered with a rather dense fulvous pubescence; head very transverse, antennal tubercle short, transverse, strongly sulcate, sulcus extending behind as far as the eyes; antennæ slender, very moniliform, first joint long, cylindrical, second ovate, third small, ovate, fourth to eighth ovate, ninth to tenth globose, eleventh shortly ovate, obtusely and somewhat abruptly acuminate; prothorax transverse, broader than the head, sides very much rounded and hardly sinuate behind the median part, lateral foveæ large, median one small and united by a strong transverse and arcuate sulcus to two minute oblong and oblique foveæ; elytra as in *F. africanus* but a little shorter. Abdomen similar.

Male: Antennæ a little longer, joints fourth to eighth more oblong, ninth to tenth globose, not transverse, eleventh ovate.

Female: Antennæ shorter, joints fourth to eighth short, ovate, ninth to tenth somewhat transverse, eleventh nearly globose. Length 1·30 mm.

This species very closely resembles *F. africanus*, for which I mistook it at first, but the antennal tubercle is shorter and more deeply sulcate, the antennæ are much shorter and much more moniliform; the prothorax is shorter, much more regularly rounded on the sides, which are not really sinuate behind, and hardly narrowed in front, so that the sides are altogether rounded from the front to the base, whilst in *africanus* the prothorax is narrowed in front, strongly rounded in the middle and sinuate towards the base; the basal impression is smaller, the elytra shorter, the colour lighter, and it is of smaller size.

Hab. Cape Colony (Cape Town, Newlands).

Much rarer than *F. africanus*.

Tribe EUPLECTINI.

Gen. TRIMIODYTES, Raffr.,

Catal., p. 52.

Trimiodytes palustris, Raffr.,

Loc. cit., p. 52.

When I described this species I had only one example at my disposal; since then I have found again this insect in the same locality. It has no sexual mark whatever on the abdomen, and what I supposed to be the female proves to be the male. The female has the head smaller and more rounded in front, the antennæ are shorter, joints fourth to eighth slightly transverse, ninth to tenth decidedly transverse, eleventh globose, truncate at base and abruptly acuminate at apex, whilst in the male the fourth to seventh joints are somewhat longer than broad, eight is square, nine to ten are very little transverse, and eleven is ovate.

TRIMIODYTES BREVIPENNIS.

Chestnut, shining; antennæ and legs testaceous, pubescence long but very sparse, with some long, erect, and scattered setæ; head about as long as broad, very little narrowed in front, and having two large foveæ and two strong sulci converging in front where they are roundly connected, vertex hardly carinate; antennæ of moderate size, the two basal joints larger, third obconical, fourth to eighth moniliform, as broad as long, ninth not much larger but transverse, tenth larger, transverse, eleventh sub-conical and very acuminate; prothorax very cordate, hardly broader, but longer, than the head, lateral foveæ strong, median one small, transverse sulcus angular in the middle; elytra short, shoulders dentate, two foveæ at the base with the dorsal sulcus well defined and extending to the median part; abdomen longer than the elytra, somewhat larger, and rounded in the middle, attenuate at apex; metasternum convex; last ventral segment large, sub-triangular.

Male: Head as long as broad, hardly attenuate in front, the anterior margin of the frontal part is thick on the sides, somewhat depressed and minutely emarginate in the middle; just under the emargination, on the epistoma, there is a little notch bearing a fovea; elytra a little longer than the prothorax, less attenuate at the base, with the shoulders more quadrate.

Female: Head not quite as long as broad, and a little attenuated in front, the anterior margin of the frontal part is rounded and altogether thick; elytra hardly longer than the prothorax, attenuate at the base, with the shoulders very oblique. Length 1·20–1·40 mm.

This new species is larger than *T. palustris*, and the elytra are of the same colour as the body, the head is smaller, the elytra are broader and shorter; in comparison with *T. setifer* the head is smaller, the foveæ and sulci much deeper, and the transverse sulcus on the front is wanting, the elytra are much shorter and attenuate towards the base, whilst in *T. setifer* the sides are nearly straight.

Hab. Cape Colony (Uitenhage).

TRIMIODYTES GRACILIS.

Elongate, rufous or testaceo-rufous, with the legs and antennæ paler, pubescence short, coarse, and scattered; head large, a little attenuate in front, and having between the eyes two foveæ and two sulci joined and rounded in front; antennæ strong, first joint quadrate, second ovate, both larger than the following ones, third subobconic, fourth to eighth moniliform, becoming a little transverse, ninth a little, tenth much larger, both transverse, eleventh large ovate, acuminate; prothorax regularly cordate, longer than broad, lateral foveæ larger than the median one, transverse sulcus faint and angular; elytra with shoulders faintly dentate, dorsal stria shorter than half the length of the elytra; metasternum convex.

Male: Head a little larger than the prothorax, less attenuate in front; eyes large; elytra much longer than broad, sides hardly rounded, not attenuate at the base, shoulders oblique and well defined; last ventral segment faintly depressed; posterior tibiæ gradually thickened towards the apex, their external margin dilated before the tip in a small, rounded lamina. Length 1·10–1·50 mm.

Female: Head a little narrower than the prothorax, more attenuate in front; elytra not much longer than broad, attenuate at the base, without well-defined shoulders; the sides are more rounded. Length 1·10–1·20 mm.

Compared to *T. palustris* this species is much more elongate, the head is comparatively smaller, and the antennæ are more clavate; it is very different from *T. setifer*, owing to the much more elongated shape and the absence of transverse sulcus in the frontal part of the head. It resembles much more *T. brevipennis*, but the sulci of the head are not so deep, and the prothorax and the elytra are longer.

Hab. Cape Colony (Port Elizabeth, Uitenhage).

TRIMIODYTES CEPHALOTES,
Plate XVIII., fig. 23.

Elongate, pale rufous, pubescence short, coarse, and sparse; head large, transverse, abruptly truncate and tri-dentate, tri-fasciculate in front, between the eyes are two small foveæ and two sulci ending in front on each side of the median spine, epistoma provided with a blunt tubercle; eyes large; antennæ elongate and slender, first joint long, sub-obconical, second ovate, both much larger than the following ones, third obconical, fourth to eighth moniliform, fifth and seventh somewhat larger, ninth larger than the preceding one, little transverse, tenth of the same shape but about twice as large, eleventh ovate, a little elongate and acuminate; prothorax slightly longer but narrower than the head, cordate, lateral foveæ strong, median one smaller, transverse sulcus slender but well defined and angular; elytra longer than broad, shoulders oblique, well marked, dorsal stria strong but stopping before the median part; metasternum convex; last ventral segment hardly impressed transversely. Male. Length 1·30 mm.

This species will be easily and at once distinguished by the shape of the head.

Hab. Cape Colony (Port Elizabeth).

GEN. EUPLECTUS, Leach.,

Catal., p. 53.

EUPLECTUS TUBERCULICEPS,

Plate XVIII., figs. 4 and 5.

Elongate, narrow, rufous; elytra, antennæ, and legs a little paler, pubescence very short, scarce and decumbent; head large, flat, little longer than broad, a little attenuate in front, sides oblique, between the eyes posteriorly, are two foveæ not as distant from each other as they are from the eyes, with a very blunt tubercle between them, two sulci slightly arcuate and not connected in front; antennæ slender, club very little distinct, first joint a little elongate, second ovate, both larger than the others, third to tenth moniliform, ninth and tenth a little more transverse and faintly larger, eleventh large, ovate; prothorax longer than broad, hardly longer and broader than the head, more attenuate in front than behind, sides rounded in the middle and posteriorly bi-sinuate but not dentate, lateral foveæ strong, median one smaller, those three foveæ connected by a strong, not much angulated, transverse sulcus, discal fovea strong, sulciform; elytra quadrate, but longer than broad, discoidal sulcus strong, not longer than the third part; fourth dorsal segment larger than the third; metasternum convex; fifth ventral segment shorter than the fourth, sixth of nearly the same size as the fourth, arcuately emarginate, seventh large, triangular, obtuse at the apex with a carina a little arcuate and asymetric. Male. Length 1·60 mm.

This species differs from all the other African ones by its more slender body, the sculpture of the head, and the absence of a spine on the sides of the prothorax behind the lateral fovea.

Hab. Cape Colony (Port Elizabeth).

GABATA, nov. gen.

This new genus, which belongs to the tribe of the *Euplectini*, is very closely allied to *Euplectus*. A reference to the description of *Euplectus* (Catal., p. 53) will be sufficient to show the differential characters. The head is smaller, and much attenuate in front, which gives it a somewhat triangular facies; the three-jointed antennal club is hardly conspicuous; the last joint of the maxillary palpi is much more elongate and fusiform; the median discoidal groove on the prothorax is wanting; the seventh ventral segment of the abdomen is very different, being small, transverse, and without carina.

With the exception of the different shape of the last joint of the maxillary palpi, which is not a very important character, the much more important difference in the structure of the seventh ventral segment of the male, the differences between this new genus and *Euplectus* consists merely in plastic modifications which might otherwise be considered as purely specific.

In a paper upon the tribe of *Euplectini*, which is being now printed, I have shown that such plastic modifications have a great value, and become generic characters, on account of their constant coincidence with important sexual modifications which are to be found in the seventh ventral segment of the male.

In *Euplectus* and some other genera this seventh segment is large, rhomboidal, and has a longitudinal carina which is nothing else but the indication of a cleavage of this segment, which opens longitudinally on both sides at the middle, to allow the extrusion of the penis; in some other genera, instead of a longitudinal cleavage, it is an operculum, which is lifted to allow the penis to protrude; in other genera this seventh ventral segment is small, more or less transverse, and hinged in such a way as to leave, when opened, between itself and the last dorsal segment, an opening for the extrusion of the penis; such is the form of the seventh ventral segment in this new genus.

I consider such sexual modifications as very important and of generic value, but unfortunately they are to be found in the male only, and it would be impossible for the females to be identified without the adjunction, as generic characters, of those plastic modifications which, by themselves, would not be sufficient to warrant the creation of a new genus.

GABATA SEMIPUNCTATA,

Plate XVIII., figs. 6 and 7.

Elongate, chestnut red, antennæ and legs paler, rufous, pubescence fine and short, decumbent and whitish; head coarsely and densely punctate, hardly as broad as the prothorax, much attenuate in front with the sides oblique, a very deep transverse channel separates the frontal part from the head, two deep and oblique sulci make an acute angle whose apex is above the transverse sulcus, and between those sulci the surface of the head is somewhat raised; behind, on the vertex, there is a faint and short longitudinal depression; antennæ short, with the basal joints much larger than the others, the first is square, second briefly ovate, third to eighth moniliform, ninth and tenth a trifle larger and transverse, eleventh much larger and ovate; prothorax finely and sparsely punctate, longer than broad, cordiform, and having three foveæ—two large lateral ones and a much smaller median one—connected by a fine transverse sulcus, and an exceedingly faint longitudinal sulcus disappearing in front; elytra impunctate, longer than broad, a little attenuate at base, with the shoulders rounded and mutic, and having two large grooves at the base and a large dorsal sulcus ending before the median part; abdomen longer than the elytra, the three first dorsal segments equal, fourth larger, first and second a little impressed at the base in the middle; under part of the head coarsely punctate; metasternum a little transverse and convex; second, third, and fourth ventral segments equal, fifth smaller, sixth as long as the fourth, depressed in the middle, with a blunt tubercle in each side on the edge, seventh small, transverse, paler than the others, densely pubescent, and with a faint longitudinal depression; tibiæ a little thickened past the middle. Male. Length 1·60 mm.

The female is not known.

Hab. Cape Colony (Port Elizabeth).

GEN. RAFFRAYIA, Reitter,

Catal., p. 62.

The number of species included in this genus, which seems to be decidedly a South African one, is constantly on the increase. In the previous Catalogue I mentioned 19 species, and to-day 26 are known, which makes it necessary for me to give a remodelled synopsis.

Synopsis of Species.

A². First dorsal segment of the abdomen much larger than the others.
  B². Antennæ slightly clavate, the penultimate joints (more especially the ninth) larger than the intermediate ones .. .. *caviceps.*
  B¹. Antennæ not clavate, the three penultimate joints (especially the ninth) smaller than the intermediate ones.
    C². Third joint of the antennæ strongly transverse .. .. .. *deplanata.*
    C¹. Third joint of the antennæ at least as long or longer than broad, triangular or globose, never transverse.
      D². Longitudinal carina of the head not extending on the frontal part, ending on the vertex.
        E². Longitudinal sulcus of the prothorax generally wanting or exceedingly faint and hardly conspicuous when it exists.
          F². Antennæ short and thick, ninth and tenth joints transverse.
            G². Elytra hardly longer than broad ; head rather long, not at all attenuate (male), little attenuate (female) ; longitudinal sulcus of the prothorax very faint and sometimes wanting *frontalis.*
            G¹. Elytra longer than broad ; head short, much attenuate in front ; longitudinal sulcus of the prothorax always entirely wanting .. .. *calcarata.*
          F¹. Antennæ much more slender, ninth and tenth joints globose, not transverse, or hardly so.
            G². Head large and thick, sides rounded, the sulci shallow, arcuate, the carina on the vertex obsolete and very short *incerta.*
            G¹. Head smaller, sides oblique, sulci deep, large and oblique, the carina on the vertex long and strong .. .. *variabilis*
        E¹. Longitudinal sulcus of the prothorax never absent, always very conspicuous.
          F². Prothorax strongly cordiform, as long or nearly as long as broad, longitudinal sulcus not very deep but very conspicuous, transverse one angulate in the middle ; shoulders generally attenuated in both sexes.
            G². Antennæ more slender, ninth joint globose, colour generally darker, piceous-brown .. .. *armata.*
            G¹. Antennæ much thicker, ninth joint transverse, colour ferruginous ; sometimes the shoulders are quadrate in both sexes .. .. *nasuta.*
          F¹. Prothorax very little cordate, broader than long, longitudinal sulcus complete and deep, transverse one straight ; shoulders very quadrate in both sexes .. .. *cruciata.*
      D¹. Longitudinal carina of the head extending from the neck to the frontal part .. .. *sulcatula.*
  A¹. First dorsal segment of the abdomen not larger than the following ones.
    B². Prothorax variable but never transversely dilated, and broader than the elytra.
      C². Head with two foveæ and two sulci.
        D². Antennæ with the joints (at least the intermediate ones) transverse.

E². Prothorax with a longitudinal sulcus more or less obsolete and sometimes reduced to an oblong fovea on the anterior part of the base.
  F². Head without any transverse sulcus on the frontal part.
    G². Prothorax transversely ovate, not cordiform .. .. *laticollis.*
    G¹. Prothorax cordiform, at least as long as broad.
      H². Longitudinal sulcus deep and well defined, head and prothorax punctate .. .. *rugosula.*
      H¹. Longitudinal sulcus more or less interrupted or obsolete; head and prothorax not punctate.
        I². Broad; antennæ with intermediate joints slightly transverse, ninth and tenth nearly quadrate; prothorax ampliated on the sides; elytra slightly longer than wide; ferruginous or testaceous .. .. *majorina.*
        I¹. More slender; antennæ with the intermediate joints and also the ninth and tenth very transverse; prothorax longer, not ampliated on the sides; elytra longer than broad; colour generally dark, feet rufous .. .. *bicolor.*
  F¹. Head with a deep transverse sulcus on the frontal part, dividing in two the tubercles bearing the antennæ; longitudinal sulcus of the prothorax faint, disappearing in front .. .. *montana.*
E¹. Prothorax without any trace of a longitudinal sulcus, the ante-basal fovea round or absent.
  F². Head without any transverse sulcus on the frontal part.
    G². Broad and convex; prothorax slightly cordate, broader than long; elytra not much longer than wide .. .. *natalensis.*
    G¹. Narrow, depressed; prothorax much cordate, longer than broad; elytra much longer than broad.
      H². Larger; head scarcely narrowed in front, sulci deep and very oblique; prothorax sinuose on the sides close to the transverse sulcus .. .. *pilosella.*
      H¹. Smaller; head strongly narrowed in front, sulci fine, little arcuated and less distant from each other; prothorax regularly cordate without sinuosity on the sides .. .. *abdominalis.*
  F¹. Head with a more or less deep transverse sulcus on the frontal part, dividing in two the tubercles bearing the antennæ.
    G². Head longer than broad, more or less attenuate in front.
      H². Ferruginous or rufous; antennæ compact and rather short; joints third to tenth transverse.

The three following species are closely allied to each other. It may be found difficult to identify the females, but the males have the following striking characters:—

        I². Head attenuate in front, sides decidedly oblique; prothorax more rounded on the sides and in front, more deeply sinuate behind the middle; male; intermediate trochanters with a basal tooth, posterior ones simple; last ventral segment with a large, oval, longitudinal and deep depression .. .. *capensis.*
        I¹. Head little attenuate in front, sides very little oblique; prothorax less rounded on the sides, attenuate in front, and less deeply sinuate behind the middle; male; intermediate

trochanters with a basal tooth, posterior ones with a small carina; last ventral segment with a large but not deep transverse depression .. .. *algoensis.*

I'. Head smaller, not attenuate in front, sides parallel; prothorax similar to that of preceding species; male; intermediate trochanters simple and mutic, posterior ones with a small, cariniform hook; last ventral segment with a smaller, superficial, transverse depression .. .. *microcephala.*

H'. Black; antennæ longer and slender, joints third to seventh only slightly transverse, eighth to tenth quadrate .. .. *obscura.*

G'. Head as broad as long, large; antennæ little compact, ninth joint quadrate, tenth very little transverse .. .. *nodosa.*

D'. Antennæ elongate, joints quadrate or even longer than broad *longula.*

C'. Head with four foveæ and without sulci .. .. *myrmecophila.*

B'. Prothorax very transverse, dilated on the sides, and wider than the elytra .. .. *dilatata.*

RAFFRAYIA FRONTALIS, n. sp.,
Plate XVIII., fig. 3.

Oblong, little convex, chestnut or testaceous, antennæ and legs testaceous, moderately pubescent, the head is variable in both sexes, but the sulcus is always rounded, and the vertex has a short carina; antennæ short and thick, first joint long, somewhat obconical, second globose, third transversely triangular, fourth to seventh much transverse, the fifth is the largest, the eighth—and especially the ninth—much smaller, transverse, tenth larger, less transverse, eleventh shortly ovate, abruptly conical at apex; prothorax cordate, a little broader than the head, sides well rounded and hardly sinuate posteriorly, longitudinal sulcus always extremely slender and sometimes wanting, the transverse one not very strong and a little angular, with the median groove small and the lateral ones a little oblong; elytra sparsely sub-rugose, short, little attenuate at the base, with the shoulders rounded, dorsal sulcus strong, reaching at least the median part; first dorsal segment very large, deeply impressed at base; metasternum convex in both sexes.

Male: Head nearly as long as broad, very little attenuate in front; frontal part large, rounded, convex, densely punctulate and squamose; eyes larger; there are no other sexual marks, even in the abdomen.

Female: Head a little shorter and more attenuate in front; frontal part truncate, not convex and smooth; antennæ somewhat thicker; elytra broader than long. Length 1·20 mm.

This species is distinct from *R. nasuta* owing to the longitudinal sulcus of the prothorax, which is hardly visible, the much shorter elytra and the shape of the head, especially in the male.

Hab. Cape Colony (Constantia, Newlands).

RAFFRAYIA SULCATULA.

Oblong, somewhat convex, ferruginous, rufous or testaceous, covered with a pale pubescence, legs and last joints of the antennæ lighter in colour; head large, shorter than broad, sides rounded, attenuate in front, foveæ and sulci very deep, a somewhat geminate and deep impression on the frontal part, and a long carina extending from the neck to the front; eyes very small; antennæ rather elongate and slender, first joint sub-cylindric, second sub-quadrate, longer than broad, third quadrate, smaller than the following one, fourth, the largest, little transverse, fourth to ninth the same form but slightly decreasing, tenth broader and more transverse, eleventh briefly ovate with the apex abruptly conical; prothorax very cordate, sides and anterior margin well rounded together, hardly broader than the head, longitudinal sulcus very feeble but never totally wanting, sides hardly sinuate posteriorly, the transverse sulcus strong, angular in the middle, and the median groove of about the same size as the lateral ones; close to the base are four small grooves; elytra smooth, much longer than broad, and very attenuate at base, no shoulders, dorsal sulcus valid, disappearing before the median part; first dorsal segment large, feebly impressed transversely at base.

Male: Metasternum hardly impressed, intermediate trochanters with a small tooth at their base, last ventral segment strongly sinuate at the apex on the sides and projecting in the middle, hardly impressed.

Female: Metasternum convex, last ventral segment rounded at apex. Length 1·70–1·80 mm.

This species is closely allied to *R. nasuta* and *R. armata*; the male will be very easily distinguished because the inferior part of the head has no sculpture and the frontal part is not produced as in *nasuta*. For the female the colour is the same as in *R. nasuta* and much lighter than in *R. armata*, the antennæ are much more slender than in both these species, the intermediate joints being hardly transverse; the size is larger.

Hab. Cape Colony (Newlands, near Cape Town).

RAFFRAYIA MONTANA.

Elongate, rufous, apex of the antennæ and legs testaceous, pubescence short and fine; head narrower than the prothorax, longer than broad, attenuate in front, and having two grooves and oblique sulci, and a well-defined transverse sulcus cutting in two the antennal tubercles; vertex feebly and shortly carinate; eyes large; antennæ thick, first joint elongate, cylindrical, second quadrate, third to ninth transverse, fourth and fifth the largest, the following ones slightly decreasing in size, tenth a little smaller and less transverse, eleventh very little larger, quadrate at base, abruptly conical at apex; prothorax cordate, about as long as broad, sinuate on the sides behind the middle; lateral foveæ large and somewhat lengthened in a fine longitudinal sulcus, median fovea small, median longitudinal sulcus feeble but never wanting, transverse one strong and angular; elytra a little longer than broad, hardly attenuate at base, with the shoulders oblique, prominent and dentate, dorsal sulcus slender and reaching the median part; first dorsal segment equal to the following one, the transverse impression at base deep, narrower than the third part, and with two short divergent carinules; metasternum more or less sulcate; last ventral segment large, transversely and feebly impressed; intermediate trochanters with a small sharp tooth in the middle. Male.

Female unknown. Length 1·40–1·60 mm.

This species resembles very much both $R.$ rugosula and $R.$ microcephala; it has, like the latter, a transverse sulcus on the front and a longitudinal sulcus on the prothorax like the first, from which it is also differentiated by the smooth teguments.

Hab. Cape Colony (Table Mountain and on the plateau above Muizenberg).

RAFFRAYIA CAPENSIS.

Elongate, ferruginous or testaceous, legs and last joints of the antennæ rufous, covered with a rather dense pubescence; head much narrower than the prothorax, a little longer than broad, attenuate, and having two large foveæ and oblique sulci on the frontal part, and a deep transverse sulcus at the base of the antennal tubercles, vertex with a very small and short carina; antennæ robust, first joint elongate, cylindrical, and quadrate, a little transverse, third to tenth transverse, decreasing a little in width from the fourth to the tenth, eleventh ovate, abruptly conical at apex; prothorax broader than long, well rounded on the sides and deeply sinuate after the middle, lateral foveæ large, median one small, longitudinal sulcus entirely wanting, transverse one strong and angular, at the base two large but not deep foveæ ; elytra a little longer than broad, not attenuate at base, shoulders oblique, little prominent, dentate, dorsal sulcus strong, and reaching the median part; first dorsal segment not larger than the other, at the base a deep impression much narrower than the third, with two very divergent carinæ ; metasternum with a small groove behind.

Male : Intermediate trochanters with a strong but short and blunt tooth at the base, posterior ones simple ; last ventral segment very large, with a deep and large oblong groove.

Female : Last ventral segment sinuate at apex, and the last dorsal one with a small tubercle. Length 1·90–2·00 mm.

This species is closely allied to *R. microcephala*, the antennæ are very much alike, but the head is not so small, and is more attenuate in front; the prothorax is broader and more deeply sinuate in the sides ; the size is larger, and the sexual characters are very different.

In *R. microcephala* I did not at first notice the presence on the posterior trochanters of a transverse, somewhat oblique, hook-like carinæ, which is very difficult to detect ; the intermediate ones are simple ; in *R. capensis* it is just the reverse, the intermediate are toothed and the posterior ones are simple.

Hab. Cape Colony (Cape Town, Kloof Road).

RAFFRAYIA ALGOENSIS.

Elongate, ferruginous, last joint of the antennæ and palpi testaceous, pubescence short and fine, intermixed with long hairs ; head longer than broad, very little attenuated in front, between the eyes two deep foveæ and two deep sulci, nearly parallel, frontal part depressed in the middle ; antennal tubercles transversely sulcate ; the vertex is transversely raised, and close to the neck there is a short carina ; antennæ not clavate, first joint elongate, second quadrate, third sub-triangular, fourth to tenth transverse, fifth the largest, eighth the smallest, ninth and tenth about the same size, eleventh hardly broader and abruptly sub-conical at tip ; prothorax much broader than the head, about as long as broad, somewhat sharply rounded on the sides at the middle, and sinuated behind by the lateral groove, which is large, slightly attenuate in front, transverse sulcus deep, angular at the middle, longitudinal sulcus entirely wanting, behind the transverse sulcus the base is convex, with four grooves; elytra broader than the prothorax, longer than broad, shoulders oblique and dentate, dorsal sulcus terminating at the median part, sides hardly rounded; first dorsal segment not larger than the following one, with two very divergent and strong carinules, including about the fourth part of the disk; metasternum longitudinally depressed; intermediate trochanters having at the base a short and recurved spine, posterior ones with a small longitudinal carinule; posterior tibiæ with a small spur; last ventral segment large and transversely depressed. Male. Length 2·10 mm.

This species is closely allied to *R. capensis* and *R. microcephala*. From *R. capensis* it differs by the smaller size, the head less attenuate in front, and the prothorax less deeply sinuated behind the middle. From *R. microcephala* it differs by the head a little attenuated, whilst in *microcephala* the sides are parallel and the head is altogether smaller.

I do not know the female of *R. algoensis*, but I think it must be very similar to the female of *R. microcephala*, and probably very difficult to distinguish. Although the females of these three species are very similar to each other, the identification of the males will not be difficult, a very frequent case in *Pselaphidæ*.

Hab. Cape Colony (Port Elizabeth).

RAFFRAYIA MYRMECOPHILA.

Plate XVIII., fig. 2.

Sub-elongate, entirely testaceous (one example, perhaps immature), covered with a white pubescence; head small, trapezoidal and transverse, frontal part somewhat depressed in the middle, between the eyes are two deep grooves, and before the front two other ones much smaller and much more closely set, no sulci; vertex carinate; eyes small; antennæ robust, second joint quadrate, third triangular, as long as broad, fourth to eighth very transverse, the fifth is the largest, and from the fifth to eighth the joints decrease in size, ninth is much narrower, transverse, tenth larger, more transverse, eleventh large, briefly ovate with the apex somewhat cone-shaped; prothorax much larger than the head, cordiform, sides rounded and made sinuose after the median part by a very strong lateral fovea, median fovea moderate, transverse sulcus not very deep and very little angular, longitudinal sulcus very faint and only conspicuous in the anterior part of the disk, base with two small foveæ; elytra a little longer than broad, very little attenuate at base, shoulders dentate, dorsal sulcus wide, reaching the middle; first dorsal segment equal to the following one, the impression at base narrow. Metasternum convex. Length 1·70 mm.

This species resembles *R. rugosula*, but differs by the absence of punctures, the head is much shorter and the cephalic foveæ are free and not connected by sulci; the longitudinal sulcus on the prothorax is exceedingly faint and may prove to be missing in other examples. The male is unknown.

Hab. Port Elizabeth.

Found with *Rhoplaomyrmex transversinodis*, Mayr., *in litt.*, a new genus of ant.

Raffrayi dilatata,
Plate XVIII., fig. 1.

Elongate and sub-parallel, more or less darkly piceous-brown, with the elytra brownish red or dark chestnut; antennæ and legs rufous, pubescence small and thin; head hardly longer than broad, attenuate in front, rounded behind the eyes, two small foveæ and two sulci converging in a median depression of the front, a very faint carinula close to the neck; antennæ slender, a little clavate, first joint somewhat short, second quadrate, following ones a trifle smaller, third to seventh quadrate, diminishing, however, in length, eighth a little smaller and transverse, ninth and tenth a little larger and transverse, eleventh larger, ovate, abruptly acuminate; prothorax more than twice wider than the head, a little broader than the elytra, transverse rounded and dilated on the sides, much narrowed behind, transverse sulcus deep, angular and widened in the middle, a faint longitudinal depression on each side and a trace of a median one, base itself with four grooves and a short median carinule; elytra much longer than the prothorax, sides a little rounded, shoulders oblique, well marked and dentate, dorsal stria a little arcuate terminating at the middle; first dorsal segment of the abdomen not larger than the following one with two strong, divergent carinæ, including nearly the third part of the disk; metasternum convex; trochanters simple; fourth ventral segment very transversely depressed, its apical margin sharp, sub-carinate, and provided on each side with a long, thin and recurved spine; last one large, thickly clothed on the sides with a golden pubescence, glabrous and depressed in the middle. Male. Length 2·00 mm.

A very curious species which differs from all the others by its broad, transverse prothorax, ampliated laterally. The female is unknown.

Hab. Cape Colony (Port Elizabeth).

RAFFRAYIA NATALENSIS, Raffray,
Catal., p. 75.

The colour varies much; the original type of the description above referred was chestnut-red. I have some suspicion that the example was not quite mature, as I have seen lately two other specimens, one from Natal and one from Port Elizabeth (Dr. Brauns), which are more or less piceous-brown, with the elytra red-brown and the legs and antennæ chestnut or rufous, which I think is the normal colouration.

GEN. DALMINA, Raffr.,
Catal., p. 78.

DALMINA ELIZABETHANA,
Plate XVIII., fig. 10.
Catal., p. 121.

When I gave the description of this species, I had only male examples for examination; since that time I received a good many specimens from Dr. Brauns, including the female, which seems to be far more abundant than the male.

I have nothing to alter in the description of the male referred to.

Female: Darker in colour, chestnut; elytra much shorter, attenuated at the base, with the sides rounded; antennæ with the three first joints as in the male, fourth larger than the third, sub-quadrate, fifth of the same shape but only a trifle larger, sixth to tenth transverse, a little narrower and slightly decreasing, eleventh as in the male, trochanters and tibiæ simple. Length 1·60–2·10 mm.

In this species the female is somewhat variable in size; in the large specimens the antennæ are thicker with the joints more transverse.

Compared with the female of *R. concolor* from Natal, there are the same differences in the head as mentioned already for the male, and the antennæ are much thicker with the joints more transverse; compared with the female of *R. globubicornis* the sixth to eighth joints of the antennæ are smaller. It differs from the female of *R.*

gratitudinis (pl. xviii., fig. 9) by the fourth and fifth joints of the antennæ being considerably larger.

According to Dr. Brauns's observations one male has been found with *Fustigerodes auriculatus*, Wasm., in the galleries of *Rhoplaomyrmex transversinodis*, Mayr., and all the other specimens, both male and female, under stones where no ants were met with.

This is a new and clear proof that it may often happen that an insect is found accidentally in ants' nest without being really myrmecophilous. The same case has been often proved for other insects.

Laphidioderus capensis, a Pselaphid, was originally discovered by my friend Mr. Péringuey, near Cape Town, inside the deep galleries of an ant, *Bothroponera pumicata*. I have taken myself a considerable number of the same insect under stones during the winter season, but I never found it in company of ants.

Another small beetle, *Microxenus laticollis*, Woll., is abundant in winter under stones, near Cape Town. I found it several times amongst ants, which did not seem to disturb it in the least, but generally this insect is found under stones where ants are not found. My opinion is that *Microxenus* is not interfered with in the least by the ants, which may come and run their galleries under the stone where it has set. Not only it is not driven away, but it seems quite unconcerned at their presence; it cannot, however, be considered a myrmecophilous insect.

Some heteromerous beetles of the genus *Tentyria, Stenosis*, and here, *Psaryphis, Aspila*, &c. and others, are often met with ants; their case does not seem to be quite similar to that of *Microxenus*. Those heteromera are very likely fond of the dejections, or vegetable or animal matter accumulated by the ants, and they are attracted to it for feeding purposes; it is more especially amongst the debris which surround the ants' nest that they are to be found.

Monoplius inflatus and *M. pinguis* are another case in point. These histeridous beetles feed and breed on and amongst the dejections of ground *Termitinæ* (*Hodotermes havilandi*), but those histeridous insects are not met with in the galleries of the Termite, and they cannot be, therefore, termed *sensu stricto*, termitophilous insects; they must be considered as living in the proximity of *Termes* and feeding exclusively on stercorarious matter produced by the *Termes*.

Quite different is the case of the *Clavigeridæ* and *Paussidæ*, which must be considered as really myrmecophilous, or at any rate myrmecobious. Both live in the very galleries of the ants, and are not to be met with anywhere else, except sometimes flying at sunset. The *Clavigeridæ* seem to be befriended and adopted by the ants, which derive some benefit from their presence amongst them, and may be really termed myrmecophilous; the *Paussidæ*, on the contrary, feed on the larvæ and pupæ of the ants, and force their presence amongst the ants by strength or intimidation by their voluntary emission of caustic gas, the contact of which appears to be much dreaded by the ants, as I have many times witnessed in Abyssinia with many different species of *Paussidæ* and ants, and more recently at Cape Town with *Paussus lineatus*, Thunb., and *Acantholepis capensis*. Those insects I call myrmecobious.

Cossyphodes and *Thorictus* are always found with the ants, either inside the galleries or sticking to the stones covering the ants' nest; but under what conditions they are living amongst ants is a thing which I do not know. If they are not myrmecophilous, they are certainly at any rate myrmecobicus.

Tribe BRYAXINI.

Gen. REICHENBACHIA, Leach,

Catal., p. 90.

REICHENBACHIA ACHILLIS, C. Schauf.,

Catal., p. 96.

This species varies to a great extent.

I have already mentioned (*loc. cit.*) a female variety from Muizenberg and Cape Town, in which the second and third dorsal segments of the abdomen are sharply spinose, whilst in the types the second dorsal segment alone is sharply mucronate. I have now another variety sent to me from Port Elizabeth by Dr. Brauns, which I name *inferior*, and both the male and female of which differ from the type by the size, a trifle smaller, a lighter-coloured body, and especially the antennæ, which are rufous instead of brown, and also by a lesser development of all the organs.

Male: The second ventral segment, has a large but not deep triangular depression, on the third and fourth there is a small transverse depression, on the last one a large but not deep rugosopunctate depression, with a smooth patch in the centre; intermediate femora not quite so thick; metasternum not so strongly impressed; the spurs of the fore and intermediate tibiæ are as in type. Length 1·70 mm.

Female : Second dorsal segment, instead of being sharply mucronate as in the type, is simply provided in the middle of the posterior margin with a blunt and faint tubercle. Length 1·50–1·60.

One female specimen is much smaller (length 1·35 mm.), and altogether piceous; the body is more elongate than is generally the case in the females; the antennæ are rufous, shorter and thicker, and there is a very faint and blunt tubercle at the apex of the second dorsal segment.

There are thus in this species two different forms of the male and three different forms of the female, which are nothing else but local varieties more or less developed.

R. achillis type, male: excavations of the ventral segments of the abdomen very deep. Muizenberg and Stellenbosch.

R. achillis type, female: second dorsal segment of the abdomen sharply mucronate. Found exclusively at Stellenbosch.

Var. *bimucronata*, female: first dorsal segment of the abdomen sometimes with a very faint, sharp tubercle, second and third dorsal segments sharply spinose. Found at Muizenberg and Cape Town, together with the male type.

Var. *inferior*, female and male: size a trifle smaller, antennæ rufous; ventral segments of the abdomen in the male with superficial impressions; the second dorsal segment in the female having simply a blunt and small tubercle.

Hab. Cape Colony (Port Elizabeth).

Such variations are very interesting, more especially the presence of two different forms of females with one form of male.

REICHENBACHIA ZAMBESIANA, Raffr.,
Reichenbachia decipiens, Raffray,
Catal., p. 92.

The name *R. decipiens* having been previously given to a species of the same genus, I propose to change it in *Reichenbachia Zambesiana*.

REICHENBACHIA SULCICORNIS, Raffray,
Catal., p. 90.

This species varies to a certain extent in size and in colour.

The colour may be ferruginous or chestnut, with the last joints of the antennæ more or less piceous, or piceous with the elytra red brown and the antennæ entirely piceous. The last joint of the antennæ in the male is also variable; it may be oblong, with a sulciform fovea or ovate and acuminate, with a much shorter fovea. Generally the last joint of the antennæ is shorter when the body is of a smaller size. Length 1·40–2·00 mm.

This species is recorded now from Bechuanaland (Vryburg), Mashunaland (Salisbury), Natal, Cape Colony (Port Elizabeth and Uitenhage).

REICHENBACHIA RIVULARIS, Raffray,
Catal., p. 129.

When I first described this species I had only one male specimen. I have received it since in large numbers from the Rev. O'Neil, from Uitenhage, and I am able to complete the description. It varies in colour, from rufous to dark chestnut; the carinules on the first dorsal segment of the abdomen are more or less distant, including from one-fifth to more than one-fourth of the disk. In the female the antennæ are somewhat shorter with all the joints a little shorter, and the eighth is decidedly transverse, the eleventh is smaller; the tibiæ have no spurs, and the tubercle at the base of the metasternum is smaller. Length 1·40–1·80 mm.

BRYAXONOMA, nov. gen.

Body stout, convex, attenuate in front; head, prothorax, and elytra entirely devoid of any fovea, sulcus, or stria; antennæ and palpi as in *Reichenbachia*; head beneath, with a strong longitudinal carina and somewhat depressed in the sides, in front of the eyes; elytra short, attenuate towards the base, no shoulders; abdomen large, margin rather narrow; first dorsal segment larger than the elytra, and without any impression; metasternum short and transverse; posterior coxæ very distant; second ventral segment very large; tarsi rather elongate, second joint sub-conical, third cylindrical and more slender; a very minute single claw.

This new genus, which belongs to the tribe of *Bryaxini*, differs much from *Reichenbachia* in general appearance, which is due to the shortness of the elytra, the very large dorsal segment of the abdomen, and the total absence of foveæ, sulcus, or stria.

BRYAXONOMA FILICEUM,
Plate XVIII., fig. 15.

Piceous, chestnut, rufous, or testaceous; palpi testaceous; legs always lighter in colour than the body; entirely covered above and beneath with a strong but rather distant punctuation, each puncture bearing a short and depressed seta; head not much convex, trapezoid, as long as broad, attenuate in front, with the sides oblique; eyes situated behind the median part of the head; antennæ having the two first joints much larger than the following ones: first subquadrate, second sub-cylindrical, longer than broad, third obconical, fourth to sixth sub-cylindrical, longer than broad, fifth somewhat larger, seventh square, eighth a little smaller and very little transverse, ninth much larger and transverse, tenth nearly double and very transverse, eleventh big, briefly ovate and acuminate; prothorax convex, a little broader than the head, a little transverse, equally attenuate in front and behind, sides rounded; elytra transverse, much broader at the apex than long, much attenuate towards the base, sides hardly rounded, a little sinuate at the external apical angle; first dorsal segment a little longer than the elytra, the following ones narrow; legs rather long; tibiæ hardly increased, but a little sinuate. Length 1·30 mm.

I have both sexes on which the penis and oviduct respectively are distinctly protruding, and yet I am unable to find any external sexual difference.

Hab. Found in roots of ferns growing along the walls of the mountain. Muizenberg, Cape Colony.

TRIBE PSELAPHINI.

GEN. PSELAPHOXYS, Raffray,
Rev. d'Ent., vol. ix., p. 137, 1890.

Oblong, much attenuate in front; head elongate, sulcate in the anterior part; maxillary palpi strong, first joint long, filiform, second shorter than the first one, conical, third small, transverse, fourth longer than the two first ones put together, fusiform, strongly papillose; antennæ strong, club tri-articulate; prothorax oblong; elytra much attenuated towards the base, ampliated behind; first dorsal segment very large, sub-triangular at the apex, with the sides broadly marginate, following ones small, immarginate, depressed, last one flat; first ventral segment hidden by a whitish glandular pubescence, second very large, third almost invisible, fourth and fifth visible on the sides only, sixth trapezoidal, depressed, surrounded laterally and behind by the last dorsal segment, which is conspicuous underneath as if it were part of the ventral segment; legs rather stout, one single claw to the tarsi; underneath the neck is covered with a thick, whitish glandular pubescence.

This genus, which belongs to the tribe of the *Pselaphinini*, is very closely allied to the genus *Pselaphus*, from which it differs chiefly by the maxillary palpi, which are much shorter, and the peculiar construction of the last ventral segments. It resembles very much the genus *Pselaphopterus*, Reitt., from Turcomania, but in the latter the last joint of the maxillary palpi is not papillose, and the last segments of the abdomen have a normal structure. From *Pselaphophus*, Raffr., which is a genus exclusively Australian, it differs by the maxillary palpi less elongate, the head narrower, the prothorax oblong, whilst it is cordiform in *Pselaphophus*.

When I first established this genus (*loc. cit.*), I considered it as being a mere sub-genus of *Pselaphus*, as well as *Pselaphophus*, but after further examination of a large number of examples, I do not hesitate to consider both as very distinct and valid genera.

The only species belonging to *Pselaphoxys* has been discovered in Abyssinia (*P. delicatulus*, Raffr.). I have just received from the Rev. O'Neil from Uitenhage two specimens which prove to be specifically identical with the Abyssinian insect, with, however, a slight difference, consisting in the colour of the setæ fringing the posterior margin of the elytra. In the Abyssinian examples such setæ are yellow, in the South African ones they are black; but I do not think that such a trifling difference should be considered a specific one, and am of opinion that both the examples from Abyssinia and South Africa belong to the same species.

Pselaphoxys delicatulus, Raffray,
Plate XVIII., figs. 16, 17, and 18.
Rev. d'Ent., 1882, p. 14.

Oblong, much attenuate in front, sanguineo-rufous, smooth and shining with some short whitish setæ; antennæ and legs rufous; head long, linear from the middle to the frontal part and sulcate, enlarged in front for the insertion of the antennæ, vertex much broader and convex; antennæ stout, first joint sub-cylindrical, second oblong, the others moniliform, ninth to tenth a little larger, sub-globose, eleventh large ovate, acuminate; prothorax oblong, as much attenuate in front as behind; elytra very much attenuate towards the base with the shoulders very oblique, a sutural stria and another dorsal stria a little arcuate, posterior margin with strong, thick and black setæ forming a brush close to the sutural angle; abdomen shorter than the elytra, first dorsal segment very large, flat, sub-triangular behind with the apex truncate, the following ones small, depressed, and altogether triangular; legs strong with the femora inflated. Length 1·70–1·90 mm.

Two examples. Sex uncertain.

Hab. Cape Colony (Uitenhage).

Tribe CTENISTINI.

Gen. SOGNORUS, Reitter,
Verh. Naturf. Ver. Brünn., xx., p. 202.

Entirely similar to the genus *Ctenistes*, and differs only by the antennæ, which are similar in both sexes; in the male the joints 3–7 are never lenticular, and the club is not formed by four very long and cylindrical joints as is the case with *Ctenistes*, but the antennæ in the males of *Sognorus* are similar to the antennæ of the females of the latter.

I confess that such a generic character is not of very great importance; in all the species known hitherto the body is shorter and stouter, and the facies really different, but in the new species here described the body has exactly the same facies as in *Ctenistes*, and the unique specimen is unquestionably a male with the antennæ of a female. This new species, which forms a transition between *Sognorus* and *Ctenistes* would lead to the conclusion that both those genera are synonymous, which conclusion will very likely be proved by further discoveries.

This genus includes all the species of North America recorded as *Ctenistes*, some Asiatic species, a European one, one from the West Coast of Africa (*simonis*, Reitter), and a new species from South Africa.

When I referred to this genus in 'Revue d'Entomologie,' 1890, p. 143, I said that it included also the Australian species. This is an error; the Australian species will form a new genus (*Ctenisophus*, Raffr.) on account of the presence of a strong infra-ocular spine which is not found in *Ctenistes* or *Sognorus*.

SOGNORUS O'NEILI,

Plate XVIII., fig. 26.

Oblong, fulvous, the squamæ are pale, fine and scattered, except on each side of the neck, at the posterior angles of the prothorax, in the posterior margins of the elytra and of the two first dorsal segments of the abdomen where they are thick and glandular; head long, a little angustate in front, between the eyes two punctures much more removed from each other than from the eyes, and in the middle a very obsolete oblong impression, in front a longitudinal sulcus extending on the antennal tubercle; eyes very large; palpi large, second joint thick and curved, third transversely pyriform, fourth transversely fusiform, those three joints produced and penicillated outwards; antennæ long, first to second joints quadrate, large, third longer, obconical, fourth to sixth ovate, longer than broad and increasing slightly in thickness, seventh to eighth a little longer, sub-cylindrical, ninth one-third longer than the preceding one, tenth hardly longer but thicker, eleventh one-third longer than the tenth, sub-cylindrical, obtusely acuminate at apex; prothorax longer than broad, sub-obconical, in the middle of the base a longitudinal impression covered with glandular pubescence; elytra much longer than broad, a little attenuate towards the base, shoulders obliquely rounded, sides nearly straight, at the base two strong foveæ, one sutural stria complete and a dorsal one disappearing behind the middle; first dorsal segment of the abdomen short, second twice as long; metasternum deeply and entirely sulcate, second ventral segment with the posterior margin a little raised and with an arcuate sinuation in the middle, altogether clothed with glandular pubescence, thin, flattened in the middle; legs long; tibiæ straight, thickened towards the apex and glabrous. Length 1·90 mm.

This species cannot be compared with *S. simonis*, Reitt., from West Africa, which is much smaller and much stouter; it resembles more the female of *Ctenistes imitator*, Reitt., but the antennæ are much thicker, and the sex of the unique specimen I have of this species is certainly a male, judging from the under side of the abdomen.

Hab. Cape Colony (Uitenhage).

Gen. CTENISTES. Reichenb.,
Catal., p. 103.

CTENISTES BRAUNSI,
Plate XVIII., fig 25.

Elongate and sub-parallel, body sparsely covered with thin ochraceous squamæ; head pyriform, rather convex, tri-foveate, antennal tubercle large and a little transverse, third joint of palpi stout, transverse, fourth sub-fusiform, slender, very transverse, appendages of moderate size; antennæ rather short and thick; prothorax a little transverse, not much narrowed in front, sides very little rounded, a short median fovea at base; elytra much longer than the prothorax and longer than broad, sides nearly parallel, dorsal stria very little arcuate; abdomen as long as the elytra, not broader, sides nearly parallel, second dorsal larger than the first, all the tibiæ straight, thickened and a little curved at the apex; metasternum sulcate.

Male: More parallel; elytra longer, shoulders more oblique and prominent; second dorsal segment only slightly longer than the first; first and second joints of antennæ larger than the others, quadrate, third obconical, longer than broad, fourth to seventh a little transverse, eighth hardly as long as the four preceding ones put together, cylindrical, ninth shorter than eighth, tenth as long as eighth but thicker towards the apex, eleventh not longer but thicker than tenth, and obtusely acuminate; metasternum more deeply sulcate; posterior tibiæ longer, somewhat angulate before the apex which is much thicker; tarsi, more especially the anterior ones, longer and more slender.

Female: Elytra a little shorter, somewhat attenuate at base with the shoulders less prominent; second dorsal segment much longer than the first; first and second joints of antennæ similar to those of the male, following ones thicker, second obconical, third to seventh nearly as long as broad, eighth a little broader, transverse, ninth larger, sub-quadrate, tenth still larger, sub-quadrate, eleventh nearly as long as the two preceding ones and obtusely acuminate; tarsi, more especially the anterior ones, short and thick. Length 2·10 mm.

This species differs very much from *C. australis*, Raffr., and *C. imitator*, Reitt., owing to the more parallel and elongate body and much thicker and shorter antennæ, the third to seventh joints of which are hardly transverse and the club is much shorter.

Hab. Cape Colony (Port Elizabeth).

Dr. Brauns found this species with a new genus of ants, *Rhoplaomyrmex transversinodis*, Mayr., *in litt.*

Tribe TYRINI.

Gen. CENTROPHTHALMUS, Schm.,
Bestr. Mon.; Psel. Prag., 1838, p. 7.

Camaldus Fairm.

Body oblong, little convex; head small, triangular, with a frontal tubercle; eyes very large with an infra-ocular spine; antennæ long and strong with a distinct club, approximate at base; palpi with the first joint inconspicuous, second elongate and clavate at the apex, third large, compressed, more or less triangular, elongate, obliquely truncate at apex, fourth much smaller, inserted at the inner angle of the third, aculeate, very sharp at the tip which is devoid of the usual appendage; prothorax more or less ovate; elytra large with a fine sutural stria and a more or less diffused and short discoidal sulcus; abdomen with a broad margin, rather short and depressed, the first dorsal segment much shorter than the following one, the two first bearing generally two longitudinal carinæ; all the trochanters, and more especially the intermediate ones, elongate with the insertion of the femur terminal, intermediate and posterior coxæ approximate; first ventral segment short and more or less concealed under the coxæ, third larger than the others; legs long and robust, tarsi elongate with two strong and equal claws.

The peculiar construction of the palpi being unique in the family will at once facilitate its identification. The genus is largely represented in Asia and in both the East and West Coasts of Africa. It extends north as far as Algeria, but it had not yet been recorded from South Africa, and the discovery of this genus in Mashunaland and in the southern part of the Colony proves once more that such countries belong to the general fauna of Africa, from which the rather isolated fauna of the Cape peninsula stands isolated.

Centrophthalmus marshalli,
Plate XVIII., figs. 20 and 21.

Oblong sub-depressed, obscure rufous, elytra brighter and redder, antennæ and legs more testaceous, pubescence long, rufous; head longer than broad, sides rounded, much attenuate in front, three equal foveæ, the posterior ones situated a little in front of the centre of the eyes, antennal tubercle nearly as long as broad, feebly sulcate, the infra-ocular spine long, sharp and straight; palpi testaceous and with erect setæ, the third joint long, not very broad, very obliquely truncate at apex with the external angle sharp, fourth rather long, sharply aculeate; antennæ long and slender, first joint cylindrical, longer than the two following ones put together but hardly thicker, second sub-quadrate, following ones a little more slender, third hardly longer than the second, third to seventh increasing a little in length, eighth nearly twice as long as seventh but hardly thicker, ninth to tenth of the same length, a little stouter, eleventh nearly three times as long as tenth, thickening from the base to the third anterior part and then attenuate and obtuse at apex; prothorax as broad as the head and eyes included, regularly ovate, lateral foveæ strong, median one antibasal, smaller than the others; elytra little convex and little narrowed towards the base with the shoulders rounded, much longer than broad, discoidal sulcus short and inconspicuous; abdomen shorter than the elytra, sub-depressed, first dorsal segment shorter by one-half than the following one; the carinæ are situated close to the sides and reach only the middle of the second segment; metasternum hardly sulcate; anterior femora thickened, tibiæ thickened in the middle, arcuate and a little sinuate, intermediate and posterior ones nearly straight; no sexual marks, but the unique specimen is very likely a male, judging from the long four-jointed club. Length 2·30 mm.

I do not know any other African species with such long and slender antennæ.

Hab. Zambesia (Salisbury).

CENTROPHTHALMUS BREVISPINA,
Plate XVIII., fig. 22.

Oblong, rufous or castaneo-rufous, pubescence long and fine, erect, yellow; head a little longer than broad, much attenuate in front, two small foveæ between the eyes and in front a longitudinal sulcus extending over the antennal tubercle; the infra-ocular spine is very small, and reduced to a thin and sharp tubercle; palpi testaceous with long, erect, whitish setæ; the third joint obconical, a little arcuate with a slightly oblique truncature at the apex, fourth inserted at about the middle of the truncature, short, thick at the base, rather abruptly aculeate at apex ; antennæ stout, first joint sub-cylindrical, second quadrate, both larger, third to sixth moniliform and a little transverse, seventh not broader but quadrate, eighth to ninth larger, tenth sub-quadrate, slightly increasing, eleventh large, briefly ovate, truncate at the base, rounded at the apex ; prothorax a little longer than broad, more attenuate in front than at the base, rounded on the sides, lateral foveæ small, median one larger; elytra with the posterior margin darker, longer than broad, a little attenuate towards the base with the shoulders oblique and little marked, two foveæ at base, the external one large and elongated in a broad but short sulcus ; first dorsal segment of the abdomen short, entirely bi-carinate, second more than twice as long as the first one ; the two carinæ are nearly complete ; metasternum convex, and with a fine stria ; all the femora, more especially the anterior ones, thickened, anterior tibiæ much thickened in the middle, arcuate, intermediate ones a little thickened towards the apex and a little curved, posterior ones straight ; no sexual mark. Length 1·80 mm.

This species, compared with the preceding one, differs at first sight by the much shorter and much thicker antennæ. In that respect it resembles very much *C. armatus*, Raffr., from Abyssinia, but in this species the infra-ocular spine is long and sharp, and the two carinæ on the second dorsal segment do not extend as far as the middle of the disk, whilst they are nearly entire in *C. brevispina*. Another African species, *C. villosulus*, Fairm., from Algeria, has the infra-ocular spine very small, but the joints of the antennal club are much longer.

Two examples. Female.

Hab. Cape Colony (Uitenhage).

Gen. PSELAPHOCERUS, Raffray.

Catal., p. 109.

Pselaphocerus amicus.

Plate XVIII., figs. 13, 14.

Resembles much *P. peringueyi*, Raffr., but the head is longer, narrower, not at all attenuate in front, with the sides parallel ; the hairs are darker, being black, except on the posterior margin of the elytra : the palpi are very much alike, the last joint being, however, more rounded externally, but the antennæ are very different.

Male: Antennæ long, first joint cylindrical, elongate, and punctate, second a little longer than broad, second to sixth the same shape and width, but increasing in length so that the sixth is nearly a third longer than the second, seventh about the same length as sixth, obconical and truncate at both ends, but the apex is very obliquely truncate, eighth the same width as sixth, transverse, ninth about twice the size of eighth, lenticular, tenth still larger, irregularly lenticular, with the under part largely foveated, eleventh large, very briefly ovate, with the under part largely excavate, the excavation transverse near the base, longitudinal on the inner side, and with a long brush of hairs before the apex; anterior trochanters with a blunt tubercle, absent on the femora, shoulders oblique, well defined. Length 3·10 mm.

Female: Antennæ a little shorter, second joint nearly square, third to fifth a little increasing in length, fifth somewhat conical, sixth shorter, seventh equal to sixth in length but sub-conical and broader at apex, eighth transverse, ninth much larger, lenticular, tenth similar to ninth, but a little larger, eleventh briefly ovate; elytra much attenuated at base, no shoulders, tubercles of the anterior, trochanters much stronger, a small tubercle on the femora of the same legs. Length 2·80–3·00 mm.

This species belongs to the first group of the genus owing to the shape of the palpi, but the seventh joint of the antennæ is much less dilated than in *P. peringueyi* and *P. diversus*, with the eighth much larger, consequently the antennæ is perfectly straight, whilst it appears somewhat angulate in these two species.

Hab. Cape Colony (Stellenbosch and Newlands).

In the descriptions of *P. peringueyi* and *P. diversus* I omitted to mention that in the male the shoulders are oblique and well marked, and wanting in the female.

PSELAPHOCERUS ACUTISPINA,
Plate XVIII., figs. 11, 12.

Stout and attenuate in front; piceous with the elytra dark red, the antennæ and legs ferruginous, or entirely flavous (presumably immature), pubescence long, dark, and mixed with fulvous hairs, palpi testaceous. Head long, narrow, a little attenuate in front, somewhat transversely depressed in front with the antennal tubercle obsoletely divided; between the eyes are two foveæ more distant from each other than from the edge; three last joints of the palpi triangular, narrowly and sharply produced outwardly in the shape of a fine appendage, the last one a little transverse; antennæ rather elongate, very different in both sexes; prothorax larger than the head, rather abruptly narrowed in front, dilated and rounded before the middle, at the median part of the sides a very transverse and sulciform fovea, clothed with whitish and glandular hairs; elytra attenuate at base with two foveæ, and a short, wide dorsal sulcus; in the anterior legs the trochanters have a very long and sharp spine and the femora a shorter, sharp spine; metasternum very little impressed.

Male: First joint of antennæ long, cylindrical, second to fifth very nearly equal to each other, longer than broad, sixth the same width, but more than half shorter, transverse, seventh very large, irregularly trapezoidal, larger at the apex where the margin is sinuate and obtusely dentate, eighth inserted at the inner angle of the preceding one, similar to the sixth but smaller, ninth not broader, much less transverse, tenth hardly longer than ninth, but three times broader, very transverse and produced externally, eleventh large, irregularly ovoid, the under part bearing a large, transverse, but not deep fovea; the spine of the trochanters is shorter, the one on the femora smaller and slender, the shoulders more angulate and prominent. Length 2·60 mm.

Female: Second, third, and fourth joints of antennæ very nearly equal to each other and a little longer than broad, fifth a good deal longer, sixth sub-quadrate, a little transverse, seventh longer and a little stouter than fifth, a little produced at the inner apical angle, eighth very transverse, ninth and tenth larger, transverse, eleventh oval; shoulders much less prominent, but still not entirely absent. Length 2·30 mm.

This species is very interesting inasmuch that by the conformation of the palpi it belongs to the second group, and by the antennæ to the first group of the genus, being a transitory form.

Hab. Cape Colony (Port Elizabeth, Uitenhage).

Sub-Family CLAVIGERIDÆ.

Gen. FUSTIGERODES, Raffray,
Catal., p. 117.

Fustigerodes auriculatus, Wasm,
Wien. Ent. Zeits., xvii., 1898, p. 98.

This species is very closely allied to *F. majusculus*, Pér.; the body is not so broad, and is more parallel, the colour is a little lighter, the head and the prothorax are more elongate, and the punctuation much less strong; the last joint of the antennæ is longer, thinner, and a little curvate; but the main difference is found in the abdominal processes; in *F. majusculus* these processes are flattened on the upper part and strongly carinate in each side; in *F. auriculatus* they are much more prominent, the upper part is rounded, convex, punctate and piliferous without any trace of carinule: the triangular tooth of the intermediate femora is a trifle smaller, and the tubercle placed before the apex inside the tibiæ of the same legs is not so strong, the size is also a little smaller. Length 2·10 mm. In his description the Rev. S. Wasmann gives 2·3 mm. as the size: very likely his typical specimen is a trifle larger than mine.

It differs from *F. capensis*, Pér., by the more elongated body, the colour much lighter, the head shorter and more parallel, the prothorax more cordate and not so much rounded, the more slender and much more regularly conical terminal joint of the antennæ, the more elongate elytra, and the form of the processes of the abdomen which are similar in both *F. majusculus* and *F. capensis*; the basal depression of the abdomen is also much larger in *F. auriculatus* than in *F. capensis*.

This interesting species described (*loc. cit.*) by the Rev. S. Wasmann has been discovered by Dr. Brauns in Port Elizabeth amongst the ant *Rhoplaomyrmex transversinodis*, Mayr. *in litt*. I have not seen type, but Dr. Brauns has kindly given me a male specimen which I have no doubt belongs to the same species as the one described by the Rev. S. Wasmann.

ADDITION.

REICHENBACHIA O'NEILI, n. sp.

Amongst the specimens of *R. rivularis* which the Rev. J. A. O'Neil sent me from Uitenhage, I noticed one or two specimens lighter in colour and differing in some respects from the type, but I considered them first as a mere variety. However, I drew the attention of the Rev. J. A. O'Neil to this fact, and later on he kindly sent me another lot of *Reichenbachia* collected together, and containing no less than 225 examples, including *R. sulcicornis*, Raffr., *R. rivularis*, Raffr., and what I considered at the time to be a variety of *rivularis*, but which I have now to consider a distinct species.

It is rather curious to note that out of 225 specimens collected together 75 proved to be *sulcicornis*, 130 *rivularis*, and 20 the new species which I name after its captor.

R. o'neili being of the same size and the same shape as *R. rivularis*, a comparative description will prove useful :—

rivularis, Raffr.	*o'neili*, n. sp.
	General coloration very much the same, but always lighter.
Antennæ: joints eight and nine a little longer than broad, tenth as long as broad, eleventh oblongo-ovate.	Antennæ: joints eight, nine and ten transverse, eleventh briefly ovate, thick.
Carinules of the first dorsal segment of the abdomen rather short and generally very divergent, including in width from one-fifth to little more than one-fourth of the disk.	Carinules of the first dorsal segment of the abdomen generally more elongate and less divergent, including in width from one-seventh to one-sixth of the disk.
Metasternum bearing, close to the intermediate coxæ, a large and blunt tubercle.	Metasternum without tubercle close to the intermediate coxæ, from which it is divided by a transverse groove.
Last ventral segment of the male with a faint impression.	Last ventral segment of the male without any trace of impression.

ODONTALGUS LONGICORNIS.

This new species is very closely allied to *O. vespertinus*, Raffr. (see Catal., p. 105), and differs by the following points:—

Colour a little lighter, being ferruginous with the disk of the elytra more or less reddish; head much more constricted in front, which makes the antennal tubercle appear much more prominent; antennæ more slender and more elongate in both sexes; dorsal segments of the abdomen neither carinate nor tuberculate on the apical edge.

Male: Antennæ—joints first and second larger than the following ones, second a little longer than broad, third longer than second, obconical, four to seven decreasing in length, so that the fourth is a little shorter than the third, seventh quadrate, eighth very transverse, ninth cylindrical, as long as the three preceding ones, broader, tenth cylindric, a trifle shorter and thicker, eleventh sub-cylindric, much longer and thicker than ninth, obtuse at tip; metasternum simply longitudinally sulcated, this sulcus filled up with whitish glandular pubescence, on each side, about at the middle, a short and carinate tubercle; ventral segments hardly longitudinally impressed.

Female: Antennæ—joints three to seven longer than in the male, the seventh being longer than broad, eighth but little transverse, ninth little larger, quadrate, tenth nearly twice larger than ninth, quadrate, eleventh nearly as long as the two preceding ones, broader, sub-cylindric, somewhat rounded at the base, obtuse at tip.

This species resembles more *tuberculatus*, Raffr., from Abyssinia, than *vespertinus*, but the colour is a little darker, the club of the antennæ in the male is much longer, and the longitudinal channel of the metasternum is complete and simple, whilst in *tuberculatus* it is divided in two parts by a transverse carina.

Hab. Cape Colony (Uitenhage).

PSELAPHOCERUS NODICORNIS, n. sp.

Ferruginous, disk of the elytra reddish, palpi testaceous, with a dense, long and rufous pubescence; head about twice longer than broad, hardly narrowed in front, the anterior edge is a little depressed in the middle, between the eyes, on the front part are two large grooves; eyes very large, situated beyond the middle; first joint of maxillary palpi rather long and conspicuous, cylindrical, second strongly and triangularly enlarged from the base to the apex, the external side rounded, and the external angle bearing on the upper surface a round impression, third of about the same size and the same shape, but with the external angle obtusely produced and the upper surface nearly totally impressed, fourth irregularly ovate, inner margin nearly straight, outer one rounded, sharply acuminate at tip, entirely impressed on the upper surface; antennæ short and stout, first joint cylindrical, not very long, second quadrate, both a little broader than the following ones, third as long as broad, fourth little transverse, fifth very large, longer than the three preceding ones together, ovate, somewhat compressed inside, finely reticulated, sixth small, very transverse, seventh about twice as large as the sixth, transverse, eighth similar to sixth, ninth and tenth much wider than seventh, lenticular, eleventh briefly ovate, truncate at the base, obtuse at apex; prothorax longer and broader than the head, rather abruptly constricted in front, two large and transverse foveæ filled up with whitish glandular pubescence; elytra long, attenuated at the base, shoulders oblique and little marked, at the base two strong foveæ filled up with whitish glandular pubescence, dorsal sulcus disappearing before the median part; anterior trochanters a little swollen and the femora with a small and blunt inner tubercle about the middle. Male. Length 2·50 mm.

According to the shape of the palpi, which have no thin and long appendages, this species should be included in the first group of the genus, but it is very different from every other. Whilst in all the other species of *Pselaphocerus* hitherto known the seventh joint of the antennæ is the largest of all, in *P. nodicornis* it is the fifth one.

Female unknown.

Hab. Cape Colony (Uitenhage).

CENTROPHTHALMUS BREVISPINA, Raffr.
(Vide supra, p. 408.)

The above description refers only to the female. Since then I received the male of this insect, which does not seem to be very rare at Uitenhage.

Male: Antennal club much longer, joints eight to ten, ovate, longer than broad, sub-equal, tenth however a little thicker at the apex, eleventh much larger, ovate, rather elongate and obtusely acuminate at apex.

PLATE XVIII.

1. Raffrayia dilatata, n. sp.
2. ,, myrmecophila, n. sp.
3. ,, frontalis, n. sp.
4. Euplectus tuberculiceps, n. sp.
5. ,, ,, last ventral segment ♂.
6. Galata semipunctata, n. sp.
7. ,, ,, last ventral segment ♂.
8. Dalmina globulicornis, Raffr., antennæ, ♂.
9. ,, elizabethana, Raffr., ,, ♂.
10. ,, gratitudinis, Raffr., ,, ♂.
11. Psclaphocerus acutispina, n. sp., antennæ, ♂.
12. ,, ,, maxillary palpus.
13. ,, amicus, n. sp., antennæ, ♂.
14. ,, ,, maxillary palpus.
15. Bryaxonoma filiceum, n. sp.
16. Psclaphoxys delicatulus, Raffr.
17. ,, ,, maxillary palpus.
18. ,, ,, last abdominal segments.
19. Pselaphus longiceps, Raffr., ,, ,, ,,
20. Centrophthalmus marshalli, n. sp.
21. ,, ,, maxillary palpus.
22. ,, brevispina, n. sp., maxillary palpus.
23. Trimiodytes cephalotes, n. sp., head ♂.
24. ,, setifer, Raffr., ,,
25. Ctenistes Braunsi, n. sp., antennæ, ♂.
26. Sognorus o'neili, n. sp., antennæ, ♂.

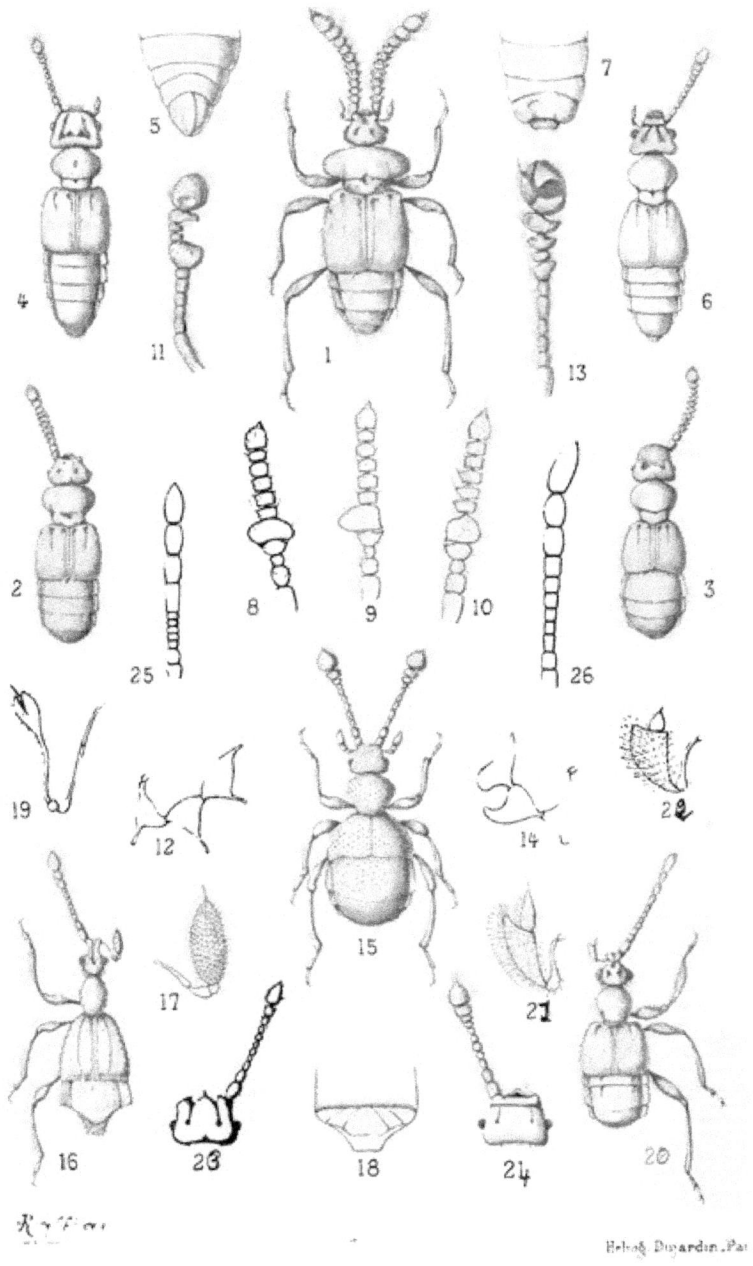

Pselaphidæ

INDEX.

A
	PAGE
achillis (Reichenbachia)	399
acutispina (Pselaphocerus)	410
algoensis (Raffrayia)	394
amicus (Pselaphocerus)	409
auriculatus (Fustigerodes)	412

B
braunsi (Ctenistes)	406
brevipennis (Trimiodytes)	384
brevispina (Centrophthalmus)	408
Bryaxonoma	401

C
capensis (Raffrayia)	393
Centrophthalmus	407
cephalotes (Trimiodytes)	385
Ctenistes	406

D
Dalmina	397
delicatulus (Pselaphoxys)	403
dilatata (Raffrayia)	396

E
elizabethana (Dalmina)	397
Euplectus	386

F
Faronidius	382
filiceum (Bryaxonoma)	402
frontalis (Raffrayia)	391
Fustigerodes	412

G
Gabata	387
gracilis (Trimiodytes)	385

L
longicornis (Odontalgus)	413

M
	PAGE
marshalli (Centrophthalmus)	407
monilis (Faronidius)	382
montana (Raffrayia)	393
myrmecophila (Raffrayia)	395

N
natalensis (Raffrayia)	397
nodicornis (Pselaphocerus)	414

O
Odontalgus	413
o'neili (Sognorus)	405
o'neili (Reichenbachia)	412

P
palustris (Trimiodytes)	383
Pselaphocerus	409
Pselaphoxys	402

R
Raffrayia	388
Reichenbachia	399
rivularis (Reichenbachia)	401

S
Sognorus	404
semipunctata (Gabata)	388
sulcatula (Raffrayia)	392
sulcicornis (Reichenbachia)	400

T
Trimiodytes	385
tuberculiceps (Euplectus)	386

Z
zambeziana (Reichenbachia)	400